Complexity in Information Theory

Yaser S. Abu-Mostafa

Complexity in Information Theory

Springer-Verlag
New York Berlin Heidelberg
London Paris Tokyo

Yaser S. Abu-Mostafa
Departments of Electrical Engineering
and Computer Science
California Institute of Technology
Pasadena, CA 91125
USA

With 7 Illustrations

Library of Congress Cataloging-in-Publication Data
Complexity in information theory.
1. Information theory. I. Abu-Mostafa, Yaser S.,
Q360.c66 1986 001.53'9 86-3746

Camera-ready copy provided by the authors.
Printed and bound by Edwards Brothers, Inc., Ann Arbor, Michigan.
Printed in the United States of America.

9 8 7 6 5 4 3 2

ISBN 0-387-96600-5 Springer-Verlag New York Berlin Heidelberg
ISBN 3-540-96600-5 Springer-Verlag Berlin Heidelberg New York

To Ayman and Samar

PREFACE

The means and ends of information theory and computational complexity have grown significantly closer over the past decade. Common analytic tools, such as combinatorial mathematics and information flow arguments, have been the cornerstone of VLSI complexity and cooperative computation. The basic assumption of limited computing resources is the premise for cryptography, where the distinction is made between available information and accessible information. Numerous other examples of common goals and tools between the two disciplines have shaped a new research category of 'information and complexity theory'.

This volume is intended to expose to the research community some of the recent significant topics along this theme. The contributions selected here are all very basic, presently active, fairly well-established, and stimulating for substantial follow-ups. This is not an encyclopedia on the subject, it is concerned only with timely contributions of sufficient coherence and promise. The styles of the six chapters cover a wide spectrum from specific mathematical results to surveys of large areas. It is hoped that the technical content and theme of this volume will help establish this general research area.

I would like to thank the authors of the chapters for contributing to this volume. I also would like to thank Ed Posner for his initiative to address this subject systematically, and Andy Fyfe and Ruth Erlanson for proofreading some of the chapters.

Y S A
Pasadena, California

FINANCIAL ACKNOWLEDGEMENT

The work of Y. Abu-Mostafa was supported by Caltech's Program in Advanced Technologies, sponsored by Aerojet General, General Motors, GTE, and TRW. The work of E. Baum was performed by Caltech's Jet Propulsion Laboratory as part of JPL's Center for Space Microelectronics Technology, and was sponsored by the Strategic Defense Initiative Organization, Innovative Science and Technology Office and the National Aeronautics and Space Administration. The work of A. El Gamal and A. Orlitsky was supported by the National Science Foundation Grant ECS-83-00988 and Air-Force Grant N0211 TASK 5. The work of J. Hopfield was supported in part by the National Science Foundation Grant PCM-84-06049. The work of J. Traub was supported by the National Science Foundation Grant MCS-82-14322 and ARPA Contract N00039-82-C-0427. The work of A. Yao was supported by the National Science Foundation Grant MCS-83-08109.

TABLE OF CONTENTS

CHAPTER I

Computational Information Theory

A. C. Yao

Department of Computer Science
Princeton University
Princeton, NJ 08544

What is information? In a fundamental sense, Shannon's definition of entropy captures the notion of information in situations where unlimited computing power is always available. As a result, in applications such as cryptography, where computational cost plays a central role, the classical information theory does not provide a totally satisfactory framework. In recent years, after Diffie and Hellman proposed the use of trapdoor function as the cornerstone for a new genre of cryptography, this deficiency is particularly dramatized; a ciphertext contains all the Shannon information about the plaintext, yet this information is 'inaccessible', i.e., it cannot be efficiently computed. This begs the challenging question 'what is accessible information?' Can we combine two very successful theories, namely, Information Theory and Computational Complexity Theory, to capture the notion of accessible information? In this chapter, we will give an exposition of a new information theory along this line and examine its applications in cryptography.

1. Introduction

In this paper we give an exposition of a new information theory and some of its applications. For a full account, we refer readers to the original paper [16].

The classical information theory originated with Shannon's 1948 paper 'A mathematical theory of communication' [13]. Althrough the central subject studied in that paper was a special communication problem, the mathematical framework developed suggested a precise definition of the amount of *information* associated with any statistical event. It is now universally recognized that Shannon's definition of information is useful and deep, as attested by the vast literature on information theory.

Then, why are we interested in a new version of information theory? There are two answers, one pragmatic and the other foundational. Firstly, there are applications such as cryptography where one would expect information theory to apply, while in fact information theory does not seem to provide a totally satisfactory framework. Secondly, on the foundational level, Shannon's definition of information is a purely statistical notion, which ignores the computational difficulty of extracting the information. In fact, the source of difficulties encountered in some applications can be directly attributed to the last point. What we seek to accomplish in the new theory is to capture the notion of *computationally accessible* information, and use it to discuss cryptography and pseudorandom number generation, subjects where information and computational complexity are closely intertwined.

In Section 2, we review Shannon's theory of cryptography and its limitations. In Section 3, we develop the central concepts of computational information theory, quantifying the notion of *computationally accessible* information. Section 4 shows that Shannon's Second Theorem and its converse are true in this new setting. In Section 5, we develop the concept of *computational security* for cryptosystems – conventional and public-key. Section 6 proposes a mathematical definition of trapdoor function. Section 7 illustrates, using the RSA function as an example, how to construct a computationally secure public-key cryptosystem from a trapdoor function. Section 8 gives an intuitive argument for the security of the resulting cryptosystem as obtained in the previous section, by demonstrating that the problem of constructing a public-key cryptosystem from a trapdoor function is the non-classical analog of Wyner's wiretap channel problem. Finally some concluding remarks are given in Section 9.

2. Motivation from Cryptography

In cryptography, the most basic problem is to design encryption methods such that two parties A and B can communicate confidential messages over public lines. In [14], Shannon developed a theory of cryptography for this problem based on his information theory. It was recognized, especially since the invention of public-key cryptosystems (Diffie and Hellman [3]), that Shannon's theory is not totally satisfactory as a foundation of cryptography. In this section, we give a brief review of Shannon's theory of cryptography and point out where its weakness lies. This motivates us to re-examine the concept of information in later sections.

Suppose A wishes to be able to send to B any message from a *message space* $W \subseteq \{0,1\}^*$. A *conventional cryptosystem* $C = (V, q, E, D)$ consists of a *key space* $V \subseteq \{0,1\}^*$, a probability distribution q over V, an *encryption function* $E : V \times W \mapsto \{0,1\}^*$, and a *decryption function* $D : V \times \{0,1\}^* \mapsto W$ such that $D(k, E(k,x)) = x$ for all $k \in V$ and $x \in W$. To use the cryptosystem, A and B agree initially on a key $k \in K$, randomly chosen according to distribution q. If A wishes to communicate to B a message (called *plaintext*) $x \in W$, A will actually send an encrypted version (called *ciphertext*) $t = E(k,x)$ to B. Clearly B can in principle *decode* the ciphertext to obtain x since $D(k, E(k,x)) = x$. An eavesdropper G, without the knowledge of k, can only infer properties of x from the ciphertext t.

Suppose the plaintext x is randomly chosen according to some distribution p over W. We wish to quantify the effectiveness of the cryptosystem against an eavesdropper. Let K, X, T denote the random variables corresponding to the values of key k, plaintext x and ciphertext t. Then the quantity $\xi(\mathcal{C}, p) = H(X) - H(X|T)$ can be regarded as the amount of information about X that the eavesdropper learns from T, where $H(X)$ and $H(X|T)$ are the standard entropy and conditional entropy functions. The smaller the ξ, the more secure $\mathcal{C}$ is for distribution p. In particular, if $\xi(\mathcal{C}, p) = 0$, i.e.,

$$H(X) = H(X|T), \tag{1}$$

the cryptosystem $\mathcal{C}$ is said to achieve *perfect secrecy* (with respect to p).

It is known that in such a case X and T are *independent random variables*, i.e., the probability distribution of (x,t) is *identical* to that of $(x, E(k,x'))$, where x and x' are two independently chosen plaintexts each distributed according to p. It also follows that the eavesdropper G cannot improve his chance of success in guessing the value of any function of the plaintext x.

Example 1. Let p be the uniform distribution on $W = \{0,1\}^m$. The *one-time pad* $\mathcal{C}_0$ is a well-known perfectly secret conventional cryptosystem with respect to p, where $\mathcal{C}_0 = (V, q, E, D,)$ with $V = \{0,1\}^m$, q being uniform on V, $E(k,x) = k \oplus x$, and $D(k,t) = k \oplus t$. (The exclusive-or represents bitwise exclusive-or.)

The main drawback of using ξ as a measure of security is that it gives too conservative an estimate. It is known that perfect secrecy implies $H(K) \geq H(X)$. Roughly, it means that the key length has to be greater than or equal to the plaintext length. In practice (see Example 2 below), one often would like to have the key length, say m', to be much smaller than the plaintext length m. But any such system is terribly insecure in this measure, since the eavesdropper can learn almost all the information about the plaintext as $\xi \approx m$.

Example 2. *Pseudo-one-time pad* $\mathcal{C}_1$.

Let m be any large integer, and $m' = \lceil m^\epsilon \rceil$, say $\epsilon = 1/4$. Let R be any pseudorandom number generator, such as linear feedback shift registers (see e.g., [9]), that takes a seed $k \in \{0,1\}^{m'}$ and outputs a pseudo-random string $R(k) \in \{0,1\}^m$. Let the message space be $W = \{0,1\}^m$, key space $V = \{0,1\}^{m'}$, encryption function $E(k,x) = R(k) \oplus x$ and decryption function $D(k,t) = R(k) \oplus t$.

If possible, we would like to have a less stringent security criterion than demanding a small ξ, such that it quarantees secrecy satisfactorily, and yet allows systems with smaller key length a chance to meet the standard.

If we apply Shannon's criterion of security to evaluate *public-key cryptosystems*, the drawback becomes obvious. Public-key cryptosystems were first proposed by Diffie and Hellman [3]. They define a *trapdoor function* as a pair $(q, \mathcal{F})$ where q is a probability distribution over $\{0,1\}^* \times \{0,1\}^*$, and $\mathcal{F}$ is a family of pairs of functions $(E_k, D_{k'})$ parametrized by key pairs (k, k') in the support of q (i.e., $q(k,k') \neq 0$), such that the following is true: (For simplicity, assume that all functions E_k, $D_{k'}$ are 1-1 mappings from $\{0,1\}^m$ to itself.)

(a) It is easy to generate a random pair (k, k') distributed according to q;

(b) $D_{k'}(E_k(x)) = x$;

(c) $E_k(x)$ is easy to compute given k and x; $D_{k'}(x)$ is easy to compute given k' and x;

(d) It is hard to compute x, given k and $E_k(x)$.

A *public-key cryptosystem* C employs such a trapdoor function as described below. B first generates in private a key pair (k, k') distributed according to q, puts the *public-key* k in the public file, and keeps k' himself as a secret *private-key*. If A wishes to send a plaintext $x \in \{0,1\}^m$ to B, A should look up k in the public file, computes $y = E_k(x)$ and send the ciphertext y to B. Clearly, B can obtain x by computing $D_{k'}(y)$. Anyone else would be unable to compute x because of property (d) above.

Let us now evaluate the security of such a system using Shannon's approach. For definiteness, let p be the uniform probability distribution over $\{0,1\}^m$, and the plaintext is randomly chosen according to p. Extending the usage of ξ earlier, let $\xi(C,p) = H(X) - H(X|K,Y)$, where K, X, Y are the random variables corresponding to the public key k, plaintext x, and ciphertext $y = E_k(x)$. The cryptosystem C achieves *perfect secrecy* with respect to p if $\xi(C,p) = 0$, i.e.

$$H(X) = H(X|K,Y). \tag{2}$$

Now $H(X) = m$ and $H(X|K,Y) = 0$ since K and Y determine the plaintext X. Thus $\xi(C,p) = m$! The system is very insecure by the ξ measure.

To reconcilliate the above analysis with the intuitive appeal of the security of the public-key system, we have to conclude that there are different kinds of information. The public key K and the ciphertext Y contain all the Shannon information concerning the plaintext X, but that information is computationally *inaccessible*! In the next three sections we develop a new kind of entropy and conditional entropy, and use them to establish a security measure for cryptosystems formally identical to the ξ measure discussed above.

Remarks. Diffie and Hellman [3] suggested that computational complexity theory might serve as a basis for future cryptography study. This line of research was taken up by a number of computer scientists (see, e.g., Blum [1], Brassard [2], Goldwasser and Micali [6][7], Rabin [11], Rivest, Shamir and Adleman [12]). The thrust of our work is to develop a theory along a different direction, in such a way as to preserve the form of Shannon's information theory of cryptography. Discussions of the relationship between these two approaches can be found in Yao [16] and Micali, Rackoff and Sloan [10].

3. Effective Entropy and Conditional Entropy

Let Γ be an *alphabet*, i.e., a finite set of symbols. Let S be a random variable, which we call a *source*, that takes on the value x with probability p_x for $x \in \Gamma$. Shannon ([13]) defined the *entropy* $H(S)$ for S as $\sum_{x\in\Gamma} p_x \log_2(\frac{1}{p_x})$. As this formula does not involve computation, it is not obvious how it should be modified when we put limits on the computing power. What we need is a description of $H(S)$ that involves computations. Shannon's First Theorem gives such a description!

Suppose two people A and B are far apart, with a source S close to A, and a noiseless communication medium by which A can send binary strings to B. A is making a series of independent observations on the values $x_1, x_2, \ldots, x_N$ of the source S, and wishes to communicate the results to B. This can be carried out with an *encoder* M_A for A which turns $(x_1, x_2, \ldots, x_N)$ into a binary string β, and a *decoder* M_B for B which does the inverse. Let L_N be the minimum average number of bits $|\beta|$ achievable by any encoder-decoder pair. Shannon's First Theorem states that $\lim_{N\to\infty} \frac{L_N}{N} = H(S)$. That is, the entropy $H(S)$ is the minimum average number of bits needed per x_i to communicate the results of observations on S.

Intuitively, the new entropy $H_c(S)$, which we will call *effective entropy*, is to be defined as the minimum average number of bits needed per x_i to communicate the results of observations on S, under the restriction that both encoding and decoding computations must be *feasible.* For a source S with a large alphabet Γ, effective entropy may be much larger than entropy. The next example illustrates why this might happen.

Example 3. Let alphabet Γ be the set of all k-bit binary strings with $k = 10^4$. For any 100-bit integers x and m, let $\alpha_{x,m}$ denote the string $c_1 c_2 \ldots c_k$ where c_j = parity of ($x^j \bmod m$). Let $\Lambda \subset \Gamma$ be the multi-set $\{\alpha_{x,m} | x, m\}$. Thus, $|\Lambda| = 2^{200}$ and $|\Gamma| = 2^{10000}$. Consider the source S over Γ with distribution density $p(y) = 1/|\Lambda|$ if $y \in \Lambda$ and $p(y) = 0$ otherwise. Clearly, $H(S) \leq \log_2 |\Lambda| = 200$. Shannon's First Theorem states that, in principle, A can send n output symbols of S to B using $nH(S) = 200n$ bits. In fact, A can simply represent each output $\alpha_{x,m}$ by the 200-bit string xm. However, in order to do this, A has to compute x and m from the 10000-bit string $\alpha_{x,m}$, as the latter is all A knows. It is not obvious that this can be done in a reasonable amount of computing time.

To search for an appropriate definition, keep in mind that we would like to model the following situation. The source S has an alphabet whose symbols are finite binary strings with an average length n (say $n \approx 200$). Person A is interested in communicating to B a sequence σ of n^t output symbols of S (t is a fixed integer, say $t = 3$). The question is, how short a string ρ can A compute in a reasonable amount of time (say in time n^k for some fixed k), so that B, on receiving ρ, can recover σ in a reasonable amount of time? To define this concept precisely, we resort to the well-developed computational complexity theory (see e.g. [8]). In this theory, the complexity of a computational problem is measured by the asymptotic behavior of algorithms for the problem as the input length becomes large. To apply the theoretical results to inputs of a particular length, we tacitly assume that this length is large enough that the asymptotic results can be used. For example, suppose theoretically one can prove that the decision problem for a certain formal logic system has complexity $\Omega(2^{2^n})$, i.e., any algorithm T for the decision problem must have a running time $\geq C_T 2^{2^n}$ for some constant $C_T > 0$. We will then regard that, for formula size $n \approx 1000$, any reasonable algorithm must use time $\approx 2^{2^{1000}}$ for some input formula. Taking this approach, we need to consider not one source, but a sequence of sources, and look at the asymptotic behavior of the quantities of interest. We now proceed to give a mathematical formulation.

Definition 1. Let Σ be a fixed, finite alphabet. A *source* S is a probability distribution p over Σ^+ with a finite expected length $\beta(S) = \sum_x p(x)|x|$. A *source ensemble* $\mathcal{S}$ is a sequence of sources $S_1, S_2, \ldots$, with probability distributions $p_1, p_2, \ldots$, such that for some fixed $t > 0$, $p_n(y) > 0$ implies $|y| < n^t$.

In the following, a *probabilistic algorithm* means a probabilistic multi-tape Turing Machine [5] that always halts. One can alternatively think of it as a program with a true random number generator available on a standard computer that always halts; our results are quite insensitive to machine models.

Notation. We will use the notation $O(v(n))$ for any function $f(n)$ that vanishes faster than $1/n^t$ for every fixed t.

The following definition specifies precisely how A is allowed to encode a sequence of n^k (for some k >0) output symbols from S_n, and how B decodes it.

Let $\mathcal{S} = \langle S_n \rangle$ be a source ensemble over an alphabet Σ, with p_n as the distribution for S_n. Let $p_n^{(k)}$ denote the distribution on $(\Sigma^+)^{n^k}$ defined by $p_n^{(k)}(\alpha) = p_n(x_1)p_n(x_2)\cdots p_n(x_{n^k})$ where $\alpha = (x_1, x_2, \ldots, x_{n^k})$.

Definition 2. A (t, k)*-encoding scheme of* $\mathcal{S}$ is a triplet of probabilisic algorithms $\mathcal{M} = (M_A, M_B, M_C)$, where M_A is a (t, k)*-encoder* from Σ to $\{0, 1\}$ and M_B is a (t, k)-decoder from $\{0, 1\}$ to Σ, satisfying the following conditions:

(a) Feed to M_A a random input (n, α), where $\alpha \in (\Sigma^+)^{n^k}$ is distributed according to $p_n^{(k)}(\alpha)$, take the output string β, and feed (n, β) as input to M_B; let γ be the ouput string of M_B; we require that $Pr\{\gamma \neq \alpha\} = O(v(n))$;

(b) Let $b > 0$ be any fixed number. Let $u = O(n^b)$, and $\alpha_1, \alpha_2, \ldots, \alpha_u$ be a sequence of strings, where each $\alpha_i \in (\Sigma^+)^{n^k}$ is independently distributed according to $p_n^{(k)}(\alpha_i)$; feed each (n, α_i) to M_A and let β_i be the output; denote by β the concatenated string $\beta_1\beta_2\cdots\beta_u$; we require that if we feed (n, β) as input to M_C, the algorithm will halt in time $O(n^c)$ for some fixed c and the output will be the u-tuple $(\beta_1, \beta_2, \ldots, \beta_u)$ with probability $1 - O(v(n))$.

For any n, define $\ell_n(\alpha)$ to be the expected length of a random output β of M_A for a random input (n, α), where α is distributed according to $p_n^{(k)}(\alpha)$. Let $\ell_n(\mathcal{M}; \mathcal{S}) = \sum_\alpha p_n^{(k)}(\alpha)\ell_n(\alpha)/n^k$, the average number of bits used by $\mathcal{M}$ to encode an output symbol x of S_n.

Remark. Property (a) stipulates that encoding and decoding can be done in polynomial computation time, and with error $O(v(n))$. Property (b) roughly states that the code is a uniquely decipherable code, in the sense that to transmit $u \cdot n^k$ outputs of the source S_n, one can encode each block of n^k outputs seperately and then concantenate the u blocks for transmission.

Definition 3. A (t,k)-*entropy sequence* for $\mathcal{S}$ is a sequence $w_1, w_2, \ldots$ such that there exists a (t,k)-encoding $\mathcal{M}$ for $\mathcal{S}$ with $\ell_n(\mathcal{M}; \mathcal{S}) = w_n$.

Only the asymptotic behavior of w_n is of interest, since for any fixed n, we can choose $\mathcal{M}$ with enough states to make $w_n = H(S_n)$, the Shannon entropy of S_n.

Definition 4. We say that the *effective entropy* $H_c(\mathcal{S}; n)$ *is less than* $g(n)$, or in symbols $H_c(\mathcal{S}; n) \leq g(n)$, if there exists $t, k > 0$ and a (t,k)-entropy sequence $\langle w_n \rangle$ for $\mathcal{S}$ such that $w_n \leq g(n)$ for all sufficiently large n.

Similarly, we write $H_c(\mathcal{S}; n) \geq h(n)$ if, for every fixed $t, k > 0$, every (t,k)-entropy sequence $\langle w_n \rangle$ for $\mathcal{S}$ satisfies $w_n \geq h(n)$ for all sufficiently large n. We will also use notations such as $H_c(\mathcal{S}; n) = O(g(n))$, $\Omega(h(n))$, $\Theta(f(n))$, etc.

Thus, we are using the term 'effective entropy of a source ensemble' in the same spirit as we speak of the 'computational complexity of a decision problem'; neither is a well-defined quantity, but can serve as a useful shordhand. In some cases, however, one has upper and lower bounds tight enough to write equalities such as $H_c(\mathcal{S}; n) = g(n) + O(\log n)$.

Example 4. The *true random number ensemble* $\mathcal{A}_0$ is the source ensemble $S_1, S_2, \ldots$ over $\{0, 1\}$ where S_n is defined to be the probability distribution $p_n(x) = 2^{-n}$ if $|x| = n$, and 0 otherwise. It is not difficult to see that $H_c(\mathcal{A}; n) = n + O(v(n))$.

Let us now try to capture the notion of *effective conditional entropy.* Consider a source ensemble $\mathcal{S} = \langle S_n \rangle$ over alphabet Σ. A *random variable* X on $\mathcal{S}$ is a sequence $\langle X_n \rangle$, where X_n is a random variable on S_n (regarded as a probability space) with values in Σ^+. Suppose the source S_n emits a sequence of n^k output symbols $\alpha_1, \alpha_2, \ldots, \alpha_{n^k}$, and person A is told the value of $X(\alpha_i)$ and $Y(\alpha_i)$ for $1 \leq i \leq n^k$, while person B is only told the value of $Y(\alpha_i)$ for $1 \leq i \leq n^k$. Now, if A wants to inform B of the value of $X(\alpha_i)$ for $1 \leq i \leq n^k$, what is the minimum average number of bits A has to send to B? Note that the question reduces to the question one faces in defining effective entropy, when Y is a constant and $X(\alpha_i) = \alpha_i$.

We will just state these definitions informally, as the details are similar to those in Definitions 2-4. A (t,k)-*encoding of* X *relative to* Y is a triplet of probabilistic algorithms $\mathcal{M} = (M_A, M_B, M_C)$, where the encoder M_A takes input $\alpha = (n, x_1, x_2, \ldots, x_{n^k}, y_1, y_2, \ldots, y_{n^k})$ and computes some binary string β, which if input together with n and $y_1, y_2, \ldots, y_{n^k}$ to the decoder M_B, will enable M_B to recover the string $x_1, x_2, \ldots, x_{n^k}$ with error probability $O(v(n))$; furthermore the codewords β are uniquely decipherable by algorithm M_C; the algorithms M_A, M_B, M_C all halt in $O(n^t)$ time. Let $\ell_n(\mathcal{M}; X|Y)$ denote the average value of $|\beta|/n^k$, that is, the average number of bits used by $\mathcal{M}$ to encode a value x of X. A (t,k)-*conditional-entropy sequence* for $X|Y$ is a sequence $\langle w_n \rangle$ such that there exists a (t,k)-encoding of X relative to Y with $\ell_n(\mathcal{M}; X|Y) = w_n$. We use the abbreviations $H_c(X_n|Y_n) \le g(n)$, $H_c(X_n|Y_n) = g(n) + O(h(n))$ etc. in the same way as in Definition 4. We will use the term *effective conditional-entropy of* $X|Y$ for $H_c(X|Y)$.

4. What Happens to Shannon's Second Theorem?

The centerpiece of information theory is Shannon's Second Theorem, which states that a channel C can reliably transmit information at a rate of R bits per channel transition provided $R < capacity(C)$. Further, a rate of $R > capacity(C)$ is impossible to achieve. The natural quesiton facing us is: How fast can a channel transmit computational information? We will see that $capacity(C)$ is still the optimum rate. In other words, computational information behaves in this context just like the classical information.

Let us first define the above question more precisely. As the details are similar to Definiton 2, we will state the definitions informally. In contrast to the way a code and its rate are defined in the Shannon case, we need to define the concepts of code and rate relative to a source ensemble.

Definition 5. Let $\mathcal{S} = \langle S_n \rangle$ be a source ensemble, and let C be a channel with input-output symbol sets I and J. A (t,k)-*coding scheme of* $\mathcal{S}$ *over* C is a triplet of probabilistic algorithms $\mathcal{M} = (M_A, M_B, M_C)$ that always halt in polynomial time $O(n^t)$. For any n and a string α of n^k outputs from S_n, the *encoder* M_A stochastically computes a string $\beta \in I^*$, sends it across channel C; the *decoder* M_B takes the resulted output string $\gamma \in J^*$ and computes a string δ. The requirement is that, when averaged over the probabilistic distribution of α and over all the stochastic moves of M_A, M_B and C, the probability of $\delta \ne \alpha$ is of order $O(v(n))$. The strings γ (output from the channel C) are uniquely decipherable by algorithm M_c (as in Definiton 2), again allowing for failure probability $O(v(n))$.

Definition 2 can be regarded as a special case of Definition 5. A (t,k)-encoding of $\mathcal{S}$ is essentially a (t,k)-coding scheme of $\mathcal{S}$ over C, where C is the binary symmetric channel with crossover probability $q = 0$.

Definition 6. In the previous definition, let $\ell_n(\mathcal{M}; \mathcal{S}; C)$ be the expected value of $|\beta|/n^k$, when the input α to M_A is generated probabilistically by source S_n.

Thus, $\ell_n(\mathcal{M}; \mathcal{S}; C)$ is the average number of channel symbols used by $\mathcal{M}$ to transmit reliably over channel C one output symbol of the source S_n. To see what 'rate', the traditional performance measure of a code, corresponds to, let us for the moment assume that $H_c(\mathcal{S}; n)$ has a sharp asymptotic behavior (e.g. $H_c(\mathcal{S}; n) = \sqrt{n} + O(n^{1/5})$). In such a case, the natural definition of the rate of $\mathcal{M}$ (for $\mathcal{S}$ over C) is

$$\begin{aligned} R_n &= \frac{H_c(\mathcal{S}; n) n^k \textit{ fbits}}{\ell_n(\mathcal{M}; \mathcal{S}; C) n^k \textit{ channel transitions}} \\ &= \frac{H_c(\mathcal{S}; n)}{\ell_n(\mathcal{M}; \mathcal{S}; C)} \textit{ fbits /channel transition.} \end{aligned} \tag{3}$$

To say that the channel capacity can be achieved means the existence of a coding scheme $\mathcal{M}$ such that $H_c(\mathcal{S}; n)/\ell_n(\mathcal{M}; \mathcal{S}; C) > (1-\delta) capacity(C)$, i.e., $\ell_n(\mathcal{M}; \mathcal{S}; C) < (1+\epsilon) H_c(\mathcal{S}; n)/capacity(C)$. Thus, the next theorem is the analogue for Shannon's Second Theorem.

Theorem 1. Let $\mathcal{S}$ be a source ensemble and C a channel. Let $\langle w_n \rangle$ be any entropy sequence for $\mathcal{S}$. Then, for any fixed $\epsilon > 0$, there exists a coding scheme $\mathcal{M}$ for $\mathcal{S}$ over C such that

$$\ell_n(\mathcal{M};\mathcal{S};C) \leq (1+\epsilon)\frac{w_n}{capacity(C)}.$$

This result is not hard to prove by a suitable use of standard results on efficient coding, e.g., Forney's concatenated codes [4]. The converse question is more interesting, namely: Can C transmit computational information reliably at rate higher than $capacity(C)$? Again, using the insight offered by (3), we see that a 'no' answer corresponds to the next theorem. This is the analogue to the *converse* of Shannon's Second Theorem.

Theorem 2. Let $\mathcal{M}$ be an encoding scheme for a source ensemble $\mathcal{S}$ over a channel C. Then for any fixed $\epsilon > 0$, there exists an entropy sequence $\langle w_n \rangle$ for $\mathcal{S}$ such that

$$w_n < (capacity(C) + \epsilon)\ell_n(\mathcal{M};\mathcal{S};C)$$

for all sufficiently large n.

Corollary. If $H_c(\mathcal{S};n) \geq h(n)$, then

$$\ell_n(\mathcal{M};\mathcal{S};C) > \frac{h(n)}{capacity(C) + \epsilon}$$

for all sufficiently large n.

The proof of Theorem 2 is quite involved. We will not get into it here. Let us mention two reasons why Theorem 2 is important for our theory from an aesthetical point of view. First, if the theorem were false, we would have a channel (say with capacity equal to 2) for which each channel input symbol can carry at most 2 bits of Shannon information, but may carry more, say 2.4 fbits, of computational information. The natural interpretation of Shannon's Second Theorem will be lost for transmitting computational informaion. We would also have to accept that, even in simple situations, not all fbits behave alike.

The second point is the there is no obvious reason why the Theorem is true. Consider the coding of a source ensemble $\mathcal{S}$ with $H_c(\mathcal{S};n) \approx \sqrt{n}$. The codewords conceivably can be so sparsely populated that if we use the same coding to transmit over a BSC with small crossover probability q, the displaced codewords after channel transmission are still widely apart. Thus, there is no purely combinatorial obstacle for the encoding to remain a valid one for a BSC, and retain a rate higher than that allowed by Theorem 2. The fact that a consistent interpretation is obtained in a new environment after complexity-type reasoning gives us confidence that our definition is on the right track.

5. New Measure of Security

In this section we will re-examine Shannon's approach to the simple cryptographic problem discussed in Section 2.

We will consider cryptosystems that contain a *scale* parameter n which ranges over all positive integers; for each value of n, a cryptosystem functions just like the systems in Section 2. Roughly, we will require the key length to be of order n, and the plaintext length to be bounded from above by a polynomial in n. The underlying philosophy is that, when we implement the cryptosystem, we will set n to be a fairly large value, as the security tends to improve as n becomes large. For large n, only strings of length bounded by polynomials of n can be manipulated, since we regard that for a computation to be feasible, the number of operations in it has to be bounded by polynomials of n.

We need some terminology before defining what a cryptosystem is, in the present context.

Definition 7. Let $\Lambda_1, \Lambda_2, \ldots, \Lambda_s$, and $\Sigma_1, \Sigma_2, \ldots, \Sigma_t$ be fixed alphabets where $s \geq 0$, $t \geq 0$. Let $P = \langle p_{n,k_1,k_2,\ldots,k_s} | n \geq 1, k_i \in \Lambda_i^+ \text{for} 1 \leq i \leq s \rangle$ be a sequence of probability distributions on $\Sigma_1^+ \times \Sigma_2^+ \times \cdots \times \Sigma_t^+$, where the support of $p_{n,k_1,k_2,\ldots,k_s}$ has only points $(x_1, x_2, \ldots, x_t)$ with all x_i of length $\Theta(n)$. We will call P a *polynomial-time distribution ensemble* on $\Sigma_1^+ \times \Sigma_2^+ \times \cdots \times \Sigma_t^+$, if there exists a probabilistic algorithm M which, given input $(n, k_1, k_2, \ldots, k_s)$, will halt in time polynomial in n, and output a $(x_1, x_2, \ldots, x_t)$ with a probability distribution $r(x_1, x_2, \ldots, x_t)$ satisfying $||r - p_{n,k_1,k_2,\ldots,k_s}|| = O(v(n))$.

Notation. For any two probability distributions q and q' over the same discrete set, let $||q - q'||$ denote $\sum_x |q(x) - q'(x)|$.

Definition 8. Let $F = \langle f_n \rangle$, where each f_n is a function from $\Lambda_1^+ \times \Lambda_2^+ \times \cdots \times \Lambda_s^+$ into $\Sigma_1^+ \times \Sigma_2^+ \times \cdots \times \Sigma_t^+$. We will call F a *polynomial-time function ensemble* on $(\Lambda_1^+ \times \Lambda_2^+ \times \cdots \times \Lambda_s^+ \mapsto \Sigma_1^+ \times \Sigma_2^+ \times \cdots \times \Sigma_t^+)$, if there exists an algorithm which, on input $(n, k_1, k_2, \ldots, k_s) \in Z \times \Lambda_1^+ \times \Lambda_2^+ \times \cdots \times \Lambda_s^+$, will halt in time polynomial in n, and output $f_n(k_1, k_2, \ldots, k_s)$. ($Z$ denotes the set of positive integers.)

We now consider conventional cryptosystems. Let $W_n = \{0,1\}^{m_n}$ where $m_n = O(n^\lambda)$ for some fixed $\lambda > 0$. We are interested in cryptosystems that can encrypt any plaintext $x \in W_n$ when the scale parameter is n.

Definition 9. A *conventional cryptosystem* C is a sequence $\langle (q_n, \hat{E}_n, \hat{D}_n) \rangle$ of triplets, where $\langle q_n \rangle$ is a polynomial-time distribution ensemble on $\{0,1\}^+$, and $\langle \hat{D}_n \rangle$, $\langle \hat{E}_n \rangle$ are both polynomial-time function ensembles on $(\{0,1\}^+ \times \{0,1\}^+, \{0,1\}^+)$. Furthermore, for all n, $\hat{D}_n(k, \hat{E}_n(k, x)) = x$ for all $k \in \text{support}(q_n)$ and $x \in W_n$.

Let $P = \langle p_n \rangle$, where each p_n is a probability distribution on W_n. How well does C perform with respect to P? To investigate this problem, let n be any given value. Consider the following stochastic process: generate a random key k according to q_n, and generate a random plaintext x according to p_n; let K_n, X_n denote the random variables corresponding to k, x. Intuitively, the amount of accessible information gained by an eavesdropper is given by $H_c(X_n) - H_c(X_n | \hat{E}_n(K_n, X_n))$. It is thus appealing to use

$$\xi_c(C, P) = H_c(X_n) - H_c(X_n | \hat{E}_n(K_n, X_n)) \tag{3}$$

as the new security measure. And the natural analogue to (2), the condition for *perfect secrecy*, is then

$$H_c(X_n) = H_c(X_n | \hat{E}_n(K_n, X_n)). \tag{4}$$

However, as we stressed before, $H_c(X_n)$ and $H_c(X_n | \hat{E}_n(K_n, X_n))$ are in general not numbers, but abbreviations for families of entropy sequences. Much care needs to be taken in interpreting formulas such as (3) and (4). For the purpose of this exposition, we will look at only one special case. Namely, when P is the ensemble $P_0 = \langle p_n^{(0)} \rangle$ where $p_n^{(0)}$ is the uniform distribution on W_n. In this case, the analogue to perfect secrecy is given below.

Definition 10. C is *computationally secure* with respect to P_0 if $H_c(X_n | \hat{E}_n(K_n, X_n)) = m_n + O(v(n))$, i.e., every conditional entropy sequence $\langle w_n \rangle$ for $X_n | \hat{E}_n(K_n, X_n)$ satisfies $w_n \geq m_n - 1/n^t$ for all fixed $t > 0$ as $n \to \infty$.

How good is this security criterion? To narrow the scope of discussion somewhat, let us assume that we are only interested in those conventional cryptosystems whose key length m' is much smaller than the plaintext length m (say m is large and $m' \leq m^{1/3}$). Ideally, a good security criterion should possess two features. First, it should not be so stringent that no scheme can meet this criterion. Second, it should be stringent enough such that some concrete assurances can be made for any cryptosystem that meets this criterion. The prefect secrecy criterion is great on the second feature but poor on the first one. How does Definition 10 fare regarding these two features?

About the first feature, there are now several conventional cryptosystems that are good candidates for being computationally secure. In fact, as we will see later, if the 'Intractability Assumption for RSA Function' is true (see Section 6), one can even have a public-key cryptosystem that is computationally secure; this implies the existence of a computationally secure conventional cryptosystem, since A and B can share each other's private keys, and use the public keys to encrypt. There are even some pseudo-one-time pads (see Example 2) that are good candidates for computationally secure conventional cryptosystems.. This is an important topic, but we will not get into it here.

We now turn to the question of the second feature. In Section 2 we mentioned that one consequence of perfect secrecy is that the plaintext X and the ciphertext T are independent random variables; another is that an eavesdropper cannot improve his chance of success in guessing the value of any function of the plaintext x. Happily, it turns out that *computational security* also implies these two consequences in an appropriate form. We will give the statement for the second consequence below (in Theorem 3). Thus it seems that the concept of computational security is a good criterion for discussing cryptosystems.

Definition 11. Let $f = \langle f_n \rangle$, where $f_n : W_n \mapsto \{0,1\}^+$. An *eavesdropper's f-guessing algorithm* is a probabilistic algorithm M which takes a pair (n, y) as input, where $n > 0$ is any integer and $y \in \{0,1\}^+$; M will halt in time polynomial in n, and output a string $z \in \{0,1\}^+$.

For the above M, consider the following experiment for any n: generate a random (k, x), where $k \in \{0,1\}^+$ and $x \in W_n$, according to the pair of probability distribution $(q_n, p_n^{(0)})$; input (n, y) to M where $y = \hat{E}_n(k, x)$, and denote the output of M by z; let $\alpha_n(M, f, C)$ be the probability that $Pr\{z = f(x)\}$. We will call $\alpha_n(M, f, C)$ the *success rate* of M against C.

Definition 12. A *non-eavesdropper's f-guessing algorithm M'* is a probabilistic algorithm which takes an integer n as input, halts in time polynomial in n, and outputs a string z. We define the *success rate* for M' in a way similar to the case for M (except that n, instead of $(n, \hat{E}_n(k, x))$, is to be input to M'), and denote it by $\beta_n(M', f, C)$.

Theorem 3. Let $f = \langle f_n \rangle$, where $f_n : W_n \mapsto \{0,1\}^+$. Let C be a computationally secure conventional cryptosystem with respect to P_0. Then, for any eavesdropper's f-guessing algorithm M, there exists a non-eavesdropper's f-guessing algorithm M' such that $\beta_n(M', f, C) > \alpha_n(M, f, C) - 1/n^t$ for all fixed t as $n \to \infty$.

We now turn to a discussion of public-key cryptosystems. In section 2, the encrypting process is a deterministic one, i.e., a given k and a plaintext determines a unique ciphertext. To allow for greater freedom in constructing cryptosystems, we will consider the more general class of public-key cryptosystems in which the encrypting process is stochastic (as defined in Goldwasser and Micali [6][7]).

Definition 13. A *transformer M* is a probabilistic algorithm which takes as input a triplet (n, k, x), where $n > 0$ is an integer and $k, x \in \{0,1\}^+$, halts in time polynomial in n, and outputs a $y \in \{0,1\}^+$. We write $(n, k, x) \Rightarrow M \Rightarrow y$, if y is a possible output when (n, k, x) is input to M.

Definition 14. A *public-key cryptosystem C* is a triplet $(\langle q_n \rangle, M_E, M_D)$, where $\langle q_n \rangle$ is a polynomial-time distribution ensemble on $\{0,1\}^+ \times \{0,1\}^+$, the *encrypter M_E* and the *decrypter M_D* are transformers, such that the following is true: For any $(k, k') \in \text{support}(q_n)$, if $(n, k, x) \Rightarrow M_E \Rightarrow y$ and $(n, k', y) \Rightarrow M_D \Rightarrow z$, then $z = x$.

It should be clear how the cryptosystem C is to be used. We now consider its security with respect to the plaintext distribution P_0.

Let $n > 0$ be an integer. Consider the following stochastic process: Generate a pair (k, k') distributed according to q_n, generate a plaintext x according to p_n; let K_n, X_n be the random variables corresponding to k, x; let Y_n denote the random variables corresponding to the output string y, when (n, k, x) is input to the encrypter M_E.

Definition 15. $\mathcal{C}$ is computationally secure with respect to P_0 if $H_c(X_n|Y_n, K_n) = m_n + o(v(n))$, i.e., every conditional entropy sequence $\langle w_n \rangle$ satisfies $w_n \geq m_n - 1/n^t$ for all fixed t as $n \to \infty$.

Note that this definition is identical to the one for the conventional cryptosystem, if we replace Y_n, K_n by $\hat{E}_n(K_n, X_n)$. The proof of Theorem 3 depends only on the relation between the random variables X_n and $\hat{E}_n(K_n, X_n)$. Thus an analogous theorem is also true for the public-key cryptosystem.

6. What Is a Trapdoor?

Trapdoor functions were informally defined in Diffie and Hellman ([3]). A suggested candidate for trapdoor function was suggested by Rivest, Shamir, and Adleman ([12]), now known as the RSA function. We will describe it in our terminology later (Example 5).

In this section, we propose a precise mathematical definition for trapdoor functions. In Section 7 we will show how every trapdoor function can be used to construct a public-key cryptosystem that is computationally secure in the sense defined in Section 5. Although the proof of security is too lengthy to be included in this paper, we will discuss in Section 8 the intuition why it is secure.

In Section 2, a trapdoor function was informally defined as a pair $(q, \mathcal{F})$ that satisfies four conditions (a)-(d). To define it rigorously, we will consider not a single pair $(q, \mathcal{F})$, but a sequence of pairs $(q_1, \mathcal{F}_1), (q_2, \mathcal{F}_2), \ldots, (q_n, \mathcal{F}_n), \ldots$, where $\mathcal{F}_n = \{(E_k^{(n)}, D_{k'}^{(n)})\}$, such that conditions (a)-(d) are met for all large n. The process of translating these conditions into the computational complexity language is straightforward, except that one has to make some decisions when (d) is translated. What is the distribution p_n of x? How strong a quantitative requirement should we put on the difficulty of inverting $E_k^{(n)}(x) = y$, when the key k has a distribution determined by q_n and x is distributed according to p_n? Regarding the first question, we will just let the distribution p_n be part of the specification of the trapdoor function, and in fact we can let p_n be dependent on the key k. In Section 2, we tacitly assume that p_n is the distribution of the plaintext x, as the simple public-key cryptosystem discussed there encrypt the plaintext directly as $E_k^{(n)}(x)$. Since we will use trapdoor functions in more sophisticated cryptosystems, we do not need p_n to reflect the plaintext distribution. For the second question, we decide on requiring that, given k distributed as induced by q_n, and x distributed according to p_n, the effective conditional entropy of the plaintext X relative to the key K and ciphertext $E_k^{(n)}(X)$ is *non-negligible*, i.e., at least $\Omega(\frac{1}{n^t})$ for some fixed $t > 0$.

In the definitions below, we will model the ensemble $\langle \mathcal{F}_n \rangle$ as one single pair of functions (f, g), by hiding n in the keys (e.g., one can append the value of n explicitly to each of the keys (k, k')). The key-pair distribution $\langle q_n \rangle$ is modeled as P_1; the distribution of x is modeled as P_2. For clarity, we have used Σ_0^+, Σ_1^+, Σ_2^+ to denote the spaces for k', k, and x. We now formally define trapdoor functions.

Let $\Sigma_0 = \Sigma_1 = \Sigma_2 = \{0, 1\}$. Let (f, g) be a pair of functions, where $f : V_f \mapsto \Sigma_2^+$, $g : V_g \mapsto \Sigma_2^+$ with $V_f \subseteq \Sigma_1^+ \times \Sigma_2^+, V_g \subseteq \Sigma_0^+ \times \Sigma_2^+$. An element $(k, k') \in \Sigma_1^+ \times \Sigma_0^+$ is said to be a *key-pair* for (f, g) if $g(k', f(k, x)) = x$ for all x such that $(k, x) \in V_f$.

Let $P_1 = \langle c_n \rangle$ be a polynomial-time ensemble on $\Sigma_1^+ \times \Sigma_0^+$, and $P_2 = \langle h_{n,k} | n \geq 1, k \in \Sigma_1^+ \rangle$ be a polynomial-time ensemble on Σ_2^+. Let $n > 0$ be any integer. Generate a random $(k, k') \in \Sigma_1^+ \times \Sigma_0^+$

according to distribution c_n, then a random $x \in \Sigma_2^*$ according to distribution $h_{n,k}$. Let $X_n^{(P_1,P_2)}$, $K_n^{(P_1,P_2)}$ denote the random variables corresponding to x, k.

Definition 16. We will say that (f,g) is a *trapdoor function* under (P_1, P_2) if the following conditions are satisfied:

(a) Any $(k,k') \in \text{Support}(P_1)$ is a key-pair for (f,g);

(b) $H_c(X_n^{(P_1,P_2)}|K_n^{(P_1,P_2)}, f(K_n^{(P_1,P_2)}, X_n^{(P_1,P_2)})) = \Omega(1/n^t)$.

We will say that (f,g) is a trapdoor function (of degree t) if there exists P_1, P_2 such that (f,g) is a trapdoor function (of degree t) under (P_1, P_2). We remark that we obtain an equivalent definition if we replace condition (b) above by the following condition: For any probabilistic algorithm which, on input $(n,k,f(k,x))$ with (k,x) distrubuted according to $(K_n^{(P_1,P_2)}, X_n^{(P_1,P_2)})$, halts in time polynomial in n and outputs w, the probability of $w \neq x$ is $\Omega(\frac{1}{n^{t'}})$ for some fixed $t' > 0$. (t' may be different from t.)

Example 5. *RSA function.*

To generate a (k,k') distributed according to c_n, we first generate two independent random n-bit primes p, q, each uniformly chosen from all n-bit primes; then choose a random $2n$-bit integer d satisfying $\gcd(d,(p-1)(q-1)) = 1$, with all such d equally likely to be selected; finally, compute a $2n$-bit integer e with $e \cdot d \equiv 1 \bmod ((p-1)(q-1))$, and let $k = (N,d), k' = (N,e)$ where $N = p \cdot q$. This defines $P_1 = \langle c_n \rangle$. The distribution $h_{n,k}$ is the uniform distribution over the set of integers $\{j | 1 \leq j < N, \gcd(j,N) = 1\}$; this defines $P_2 = \langle h_{n,k} | n \geq k \rangle$. The function f and g are defined as $f(k,x) = x^d \bmod N$ and $g(k',x) = x^e \bmod N$.

The RSA function is an attractive candidate for trapdoor function because of its connection with integer factoring and its elegance. So far it has resisted efforts to invert the function $f(k,x) = y$. For later reference, we state the obvious conjecture in the form of a hypothesis. Let (P_1, P_2) be as in Example 3.

The Intractability Assumption of RSA. *The RSA function is a trapdoor function under* (P_1, P_2).

For later use, let t_{RSA} denote the smallest positive integer such that RSA is a trapdoor function of degree t_{RSA}.

7. Construction of a Secure Cryptosystem

We will describe the construction of a public-key cryptosystem from any candidate for a trapdoor function. We illustrate the construction by using the RSA function. There are two integer parameters n and t. The parameter $t > 0$ is a *security* parameter, whose role becomes clear later; n is the scale parameter for the system.

First we need to specify how a pair of keys (k,k') is generated. We will generate $\ell = n^t$ independent RSA key pairs. That is, for each $1 \leq i \leq n^t$, generate two random n-bit primes p_i, q_i, a random $2n$-bit integer d_i with $\gcd(d_i, (p_i - 1, q_i - 1)) = 1$, and compute e_i with $e_i \cdot d_i \equiv 1 \bmod ((p_i - 1, q_i - 1))$; now let the public key k be $(N_1, N_2, \ldots, N_\ell, d_1, d_2, \ldots, d_\ell)$, where $N_i = p_i \cdot q_i$, and the private key k' be $(N_1, N_2, \ldots, N_\ell, e_1, e_2, \ldots, e_\ell)$.

Second, we describe how the encrypter M_E computes a ciphertext, when a plaintext x is given. Note that the ciphertext is probabilistically chosen. We will encrypt each bit of x sequentially and independently, and concatenate the results. Thus, we need only specify how to encrypt a single bit $b \in \{0,1\}$. We will do this in two steps.

Step 1. In time polynomial in n, we generate $n\ell$ integers $z_i, 1 \leq i \leq n\ell$, such that

(a) x_i are randomly and independently chosen from the set of integers $I_i = \{j | 1 \leq j \leq N_i, \gcd(j, N_i) = 1\}$,

(b) b equals the mod 2 sum $\oplus_{1 \leq j \leq n} \oplus_{\ell(j-1) \leq i \leq \ell j} z_{ij}$, where z_{ij} is the j-th bit of x_i.

Step 2. Compute $y_i = z_i^{d_i} \bmod N_i$ and let the concatenated string $y = y_1 y_2 \cdots y_{n\ell} \in \{0,1\}^{n^2\ell}$ be the ciphertext for the bit b.

Lastly, we need to specify how the decrypter M_D recovers b. This is simple, since with the private key k', one can obtain z_i by computing $y_i^{d_i} \bmod N_i$, and then compute b by summing modulo 2 the appropriate bits of z_i.

Let P_0 be the uniform distribution ensemble defined in Section 5.

Theorem 4. If the Intractability Assumption for RSA is true, then C_3 is computationally secure with respect to P_0 with $t = 3(1 + t_{RSA})$.

Remark. In fact, C_3 is computationally secure with respect to any P, not just P_0. But we have restricted our discussion to P_0 only in Section 5; for general P, more care needs to be exercised in the definition of security.

8. What Makes the Cryptosystem Secure?

In this section we give an intuitive proof of Theorem 4 by resorting to connections to a problem in classical information theory. This also furnishes an illustration that, by keeping track of the computational information in the various strings, one can manipulate the computational information flow just as in the case of the classical information.

Let (f, g) be a trapdoor function under (P_1, P_2). By definition, $H_C(X_n | K_n, f(K_n, X_n)) = \Omega(1/n^t)$, where we have written X_n, K_n for $X_n^{(P_1,P_2)}$, $K_n^{(P_1,P_2)}$. For definiteness, let us assume that f maps a pair of n-bit strings into an n-bit string.

Let S_n, T_n denote the random variables (K_n, X_n), $(K_n, f(K_n, X_n))$ respectively. Then

$$H_C(S_n | T_n) = \Omega(1/n^t). \tag{5}$$

Note that the value of T_n is determined by S_n (and vice versa). In fact, let $\langle F_n \rangle$ be a sequence of functions defined by $F_n(k, x) = (k, f(k, x))$, then $T_n = F_n(S_n)$.

For any given large n, F_n can be viewed as a channel C_n with *huge* alphabets, in fact with 2^{2n} input symbols and 2^{2n} output symbols; the input symbol (k, x) is distributed according to S_n. This is a subject we have not encountered, and it is no longer true that all the conclusions based on Shannon's information theory are applicable, if we have available only a limited computing capability. Thus F_n is a *non-classical* channel, and we have to consider accessible information instead of Shannon information.

Let us imagine the situation when party A is on the input side of the channel F_n, sending string (k, x) across the channel to party G, who receives the string $F_n(k, x)$. Clearly, the Shannon conditional entropies $H(T_n | S_n)$ and $H(S_n | T_n)$ are both 0, and classically F_n is a noiseless channel. However, when we consider accessible information, we find the $H_c(T_n | S_n)$ is still 0, but $H_c(S_n | T_n)$ is $\Omega(1/n^t)$ according to (5). That is, A has no uncertainty about what G receives, but G has at least some uncertaity about what A has sent, admittedly the uncertainty may be as small as $1/n^t$ fbit. Thus, from $G's$ standpoint, he is on the receiving end of a noisy channel with a noise at least $1/n^t$, a non-neglible quantity in polynomial-time calculations.

Let us now visualize the above picture in a slightly different way. Assume that a third party B is the owner of a secret key k', while A is really transmitting x to B as $F_n(k, x)$. It is a clear channel to B, since he has the secret key to decode $F_n(k, x)$. The role of G now is an eavesdropper, who, without the secret key, is trying to wiretap the line with a low-grade equipment. But this situation has an exact analogue in the classical information theory, known as the Wyner wiretap channel [15]. Wyner showed that even when the noise in $G's$ channel is small, A can magnify the noise by properly encoding his messages. For example, suppose G has a binary symmetric channel with crossover probability 10^{-4}, and A encodes a bit b as a random 10^6-bit string α such that α has an even number of 1's if and only if $b = 0$. Then, of the transmitted bits, G can guess the value of each individual bit with high confidence, yet G knows that he is going to be wrong in the values of about 100 bits. It becomes difficult to estimate whether he has missed an even number of bits or an odd number. Indeed it was shown that schemes of the kind indeed will completely baffle G about the true value of b. For channels with larger alphabets, more sophisticated schemes are needed. We leave it for the reader to verify that our encrypting method in Section 7 has a counterpart as a solution to the wiretap channel.

In our case, the channel noise is somewhat like n^{-t}, and the approximate estimate on the length of α is about n^t, which is large but can be accomplished in polynomial time.

9. Concluding Remarks

In the last section, we saw that a trapdoor function can be viewed as a *non-classical* channel, i.e., a channel whose behavior differs from what is expected from Shannon's information theory when limitation is put on the computing power; also, we saw how such views can sometimes help in the design of algorithms, at least in the field of cryptography. One may wonder whether there is any interesting application when a source is so huge as to become non-classical. The answer is indeed yes. Pseudorandom number generation, which has so far no satisfactory theoretical foundation, has made progress towards acquiring one using computational information theory. A pseudorandom number generator is of high quality exactly when it become a non-classical source! We refer the readers to [16].

The study of accessible information is still in its infancy. There are at least two major directions for future research. One is to examine, within information theory, whether and how the consideration of computing power changes the vast number of existing results. Another is to seek applications in other fields where both the *amount* and *computability* of information play an important role. We are optimistic that both avenues will lead to fruitful discoveries.

References

[1] M. Blum, "Coin-flipping by telephone: a protocol for solving impossible problems," *Proc. IEEE COMPCON SPRING*, 1982, 133-137.

[2] G. Brassard, "Relativized cryptography," *Proc. 20th IEEE Symp. on Foundations of Computer Science,* 1979, 383-391.

[3] W. Diffie and M, E, Hellman, "New directions in cryptography," *IEEE Trans. on Inform. Theory* IT-22, 6(1976), 644-654.

[4] G. D. Forney, Jr.,*Concatenated Codes,* MIT Press, 1966.

[5] J. Gill, "Computational complexity of probabilistic Turing machines," *SIAM J. on Computing* 6(1977),

675-695.

[6] S. Goldwasser and S. Micali, "Probabilistic encryption and how to play mental poker keeping secret all partial information," *Proc. 14th ACM Symp. on Theory of Computing,* San Francisco, May 1982.

[7] S. Goldwasser and S. Micali, "Probabilistic encryption," *Journal of Computer and Systems Sciences* 28 (1984), 270-299.

[8] J. E. Hopcroft and J. D. Ullman, *Introduction to Automata Theory, Languages, and Computation,* Addison-Wesley, Reading, Mass., 1979.

[9] F. J. MacWilliams and N. J. A. Sloane, "Pseudo-random sequences and arrays," *Proceedings of IEEE,* 64 (1976), 1715-1729.

[10] S. Micali, C. Rackoff, and B. Sloan, "The notion of security for probabilistic cryptosystems," to appear in *SIAM J. on Computing.*

[11] M. O. Rabin, "Digitalized signatures and public-key functions as intractable as factorization," MIT/LSC/TR-212, 1979.

[12] R. L. Rivest, A Shamir and L. Adleman, "A method for obtaining digital signatures and public-key cryptosystems," *Communications of ACM* 21 (1978), 120-126.

[13] C. E. Shannon, "A mathematical theory of communication," *Bell System Tehnical Journal* 27 (1948), Part I, 479-523, Part II, 623-656.

[14] C. E. Shannon, "Communication theory of secrecy systems," *Bell System Tehnical Journal* 28 (1949), 656-715.

[15] A. D. Wyner, "The wire-tap channel," *Bell System Tehnical Journal* 54 (1975), 1355-1387.

[16] A. C. Yao, "Theory and applications of trapdoor functions (extended abstract)," *Proceedings 23rd IEEE Symposium on Foundations of Computer Science,* 1982, 80-91. Full paper will be available from the author on request.

CHAPTER II

Communication Complexity

A. Orlitsky and A. El Gamal

Information Systems Laboratory
Department of Electrical Engineering
Stanford University
Stanford, CA 93405

The communication complexity of a function $f:\{0,\cdots,n-1\}\times\{0,\cdots,n-1\} \rightarrow \{0,1\}$ is the number of bits two persons have to exchange in order to determine $f(x,y)$ when, initially, one person knows x and the other knows y. The following measures of complexity were considered in the literature:

* worst case / average,
* deterministic / randomized,
* error free / ϵ-error allowed.

Techniques to bound these measures are described and applied to prove asymptotically tight bounds on the complexities of 'most functions' and of some specific examples. Several extensions of the problem are detailed.

1 Introduction

Communication complexity, first introduced in [Yao 79][1], is concerned with different aspects of the following problem.

> A boolean function f is defined on a finite product set $S_{\mathcal{X}} \times S_{\mathcal{Y}}$. Two persons, $P_{\mathcal{X}}$ having a random variable $X \in S_{\mathcal{X}}$ and $P_{\mathcal{Y}}$ having a random variable $Y \in S_{\mathcal{Y}}$, want to determine $f(X,Y)$. How many bits of information do they need to exchange?

For simplicity, assume that $S_{\mathcal{X}} = S_{\mathcal{Y}} = \{0,\ldots,n-1\}$. Clearly, $P_{\mathcal{X}}$ and $P_{\mathcal{Y}}$ need never exchange more than $\lceil \log n \rceil + 1$ bits[2]: using $\lceil \log n \rceil$ bits, $P_{\mathcal{X}}$ transmits the binary representation of X, and then $P_{\mathcal{Y}}$, knowing X and Y, computes and transmits $f(X,Y)$. For some functions much fewer bits suffice: if $f(x,y) \stackrel{\text{def}}{=} (x+y) \bmod 2$, $P_{\mathcal{X}}$ can transmit $X \bmod 2$ and $P_{\mathcal{Y}}$ can transmit $Y \bmod 2$, two bits in all.

An assignment of values to X and Y is called an *input*. We assume that the *communicators* ($P_{\mathcal{X}}$ and $P_{\mathcal{Y}}$) agree a priori on a *protocol*: a set of rules that, for each input, determines (1) a sequence of transmitted bits and (2) a *computed value* that they assume is the correct value of $f(X,Y)$ for that input.

There are different types of protocols and different ways to measure the number of transmitted bits. They are illustrated in this section using the following example. $P_{\mathcal{X}}$ and $P_{\mathcal{Y}}$ want to check whether $X = Y$, i.e., f is the *equality function* ([Yao 79]):

$$equ(x,y) \stackrel{\text{def}}{=} \begin{cases} 1 & \text{if } x = y \\ 0 & \text{if } x \neq y. \end{cases}$$

Let $x_1 \cdots x_{\lceil \log n \rceil}$ denote the binary representation of an integer $x \in \{0,..,n-1\}$ and consider the following protocols.

Protocol ϕ_1:

1. $P_{\mathcal{X}}$ transmits $X_1, \ldots, X_{\lceil \log n \rceil}$.

2. $P_{\mathcal{Y}}$, knowing X and Y, transmits 1 if $X = Y$ and 0, otherwise.

---

[1] A similar model, for continuous functions, was formulated in [Ab 78].

[2] Throughout this manuscript, log denotes logarithm to the base 2.

3. They take the bit transmitted by $P_{\mathcal{Y}}$ to be the computed value.

Protocol ϕ_2:

1. Starting with $i = 1$, and continuing with consecutive i's, $P_{\mathcal{X}}$ and $P_{\mathcal{Y}}$ exchange X_i and Y_i until $X_i \neq Y_i$ or $i = \lceil \log n \rceil$.

2. The computed value is 1 if the last two transmitted bits were identical and 0 otherwise.

Both protocols are *error free for the equality function*: their computed value is equal to *equ* for all inputs. For each input, the first protocol, ϕ_1, determines a sequence of $\lceil \log n \rceil + 1$ transmitted bits. Therefore, it requires transmission of $\lceil \log n \rceil + 1$ bits both in the worst case and on the average (over all n^2 possible inputs, assumed equally likely). On the other hand, ϕ_2 requires $2\lceil \log n \rceil$ bits in the worst case (whenever $X = Y$), but only 4 bits on the average (for half the inputs, communication stops after exchanging the first pair of bits, for quarter, after the second, and so on).

It is therefore natural to ask whether there is an error-free protocol for the equality function that requires less than $\lceil \log n \rceil + 1$ bits in the worst case. The following argument, called a "crossing-sequence" argument in [LS 81], and a "fooling-set" argument in [Ul 84], provides a negative answer.

For $x, y \in \{0,..,n-1\}$, we denote the input $X = x$, $Y = y$ by (x, y). It is easy to see that if i and j are distinct integers in $\{0,..,n-1\}$, and a protocol dictates the same sequence of transmitted bits for the inputs (i, i) and (j, j), then this sequence is also transmitted for the input (i, j). But $P_{\mathcal{X}}$'s decision on the computed value is based on X and the transmitted sequence; it must therefore be the same for (i, i) and for (i, j). Hence the computed value will differ from $equ(X, Y)$ for at least one of these inputs.

Consequently, any error-free protocol for the equality function, must associate a different sequence of transmitted bits with each input in $\{(0,0), \ldots, (n-1, n-1)\}$. A more detailed treatment of the problem, provided in Section 2, shows that the set of possible sequences must be prefix free. Hence any error-free protocol must transmit at least $\lceil \log n \rceil$ bits for some input in $\{(0,0), \ldots, (n-1, n-1)\}$.

Suppose, now, that we let $P_{\mathcal{X}}$ and $P_{\mathcal{Y}}$ incur some errors in deciding on the value of $equ(X, Y)$. There are several ways to define the error measure. Perhaps the most natural, is as the proportion of inputs for which $P_{\mathcal{X}}$ and $P_{\mathcal{Y}}$ make the wrong decision. However, $equ(X,Y) = 1$ for only n of the n^2 inputs. Hence, $P_{\mathcal{X}}$ and $P_{\mathcal{Y}}$ can always decide that $equ(X,Y) = 0$ thereby incurring an error of $1/n$; as n increases, the error tends to 0.

This trivial solution, leads to a different definition of the error measure which is best understood in the context of ***randomized protocols.*** For each input, protocols ϕ_1 and ϕ_2 always result in the same sequence of transmitted bits and the same computed value. Such ***deterministic protocols*** are either always correct for a given input or always in error. ***Randomized protocols***, on the other hand, allow $P_{\mathcal{X}}$ and $P_{\mathcal{Y}}$ to base their transmissions and computed value on the outcomes of random experiments. Therefore the computed value assigned to each input is a $0-1$ random variable. We define the ***error a protocol incurs in computing*** f to be the maximum, over all inputs, of the probability that the computed value assigned by the protocol differs from the correct value of f for that input.

Clearly, any *deterministic* protocol that computes a function with at most ϵ error ($\epsilon < 1$) under this definition must be error free for that function. For the equality function this implies transmission of at least $\log n$ bits in the worst case. The next randomized protocol, due to [RY 79], is based on the "prime numbers algorithm" of [Fr 77]. It incurs at most ϵ error while transmitting only $O(\log\log n + \log\frac{1}{\epsilon})$ bits.

Define $\sigma \stackrel{\text{def}}{=} \sqrt{2}\frac{\ln n}{\epsilon}$ and let $(x)_\alpha$ denote $x \bmod \alpha$. The protocol proceeds as follows:

1. $P_{\mathcal{Y}}$ picks at random a prime α such that $\sigma \le \alpha \le 2\sigma$. He transmits α and $(Y)_\alpha$.

2. $P_{\mathcal{X}}$ transmits 1 if $(X)_\alpha = (Y)_\alpha$ and 0 otherwise.

3. They take the bit transmitted by $P_{\mathcal{X}}$ to be the computed value.

For all inputs (x,y) such that $x = y$, the computed value is always $1 = equ(x,y)$. For any input $x \neq y$ the computed value is wrong exactly when α divides $x-y$. The number of primes between σ and 2σ is $\ge \frac{\sigma}{\sqrt{2}\ln\sigma}$, and at most $\log_\sigma n$ of them divide $x-y$. Therefore, for all inputs, the probability of a wrong decision is $\le (\log_\sigma n)/\frac{\sigma}{\sqrt{2}\ln\sigma} = \epsilon$. The number of bits transmitted for any input and any selected prime is at most $2(\log\ln n + \log\frac{1}{\epsilon} + 1.11)$. For large enough values of n, this is much smaller than $\log n + 1$.

We introduced randomized protocols in the context of saving bits when ϵ maximum error is allowed. Indeed, we will see in Example 7 that for the equality function, randomization cannot help reduce the number of transmitted bits in error-free computation.

However, for some functions, this can be done: Example 6 describes a function for which every error-free deterministic protocol requires $\log n$ bits in the worst case but there exists an error-free randomized protocol for that function which, on the average, requires less than $\sqrt{\log n}\sqrt{\log\log n}$ bits for each input.

In summary, we saw that the following properties hold for all protocols computing the equality function. Every error-free deterministic protocol must transmit at least $\log n$ bits in the worst case. This is the also the case for error-free randomized protocols or ϵ-error deterministic protocols. However, when both randomization and errors are allowed, or when the average number of bits is considered, much fewer bits suffice.

In the rest of the paper we bound the number of transmitted bits needed for "most" functions, and for some specific ones. In the next section, we consider deterministic protocols. In Section 3, we analyze randomized protocols. In Section 4, we summarize and compare the results concerning most functions. In the last section, we survey other related results not treated here.

2 Deterministic Protocols

In this section we consider deterministic protocols which we formally define in Section 2.1. In Sections 2.2 and 2.3 we describe lower-bound techniques for deterministic protocols and apply them to obtain tight bounds for two functions and to show that although "most" functions require the exchange of almost $\log n$ bits both on the average and at worst, most *sparse* functions require the exchange of much fewer bits on the average than at worst.

2.1 Definition and Preliminary Properties

The definition is similar to that of [PpS 82] and [EO 84]. It differs from [PpS 82] in that (1) it views the protocol as a codebook — an approach that facilitates information-theoretic treatment, and (2) the computed value of the function does not have to be transmitted explicitly.

We assume that the communication is performed over a binary channel that can carry only one bit at a time. Therefore, each communicator, in his turn, transmits a (possibly empty[3]) string of bits that we call a *message*. For example, in protocol ϕ_1 of the introduction, the first message is $\log n$ bits long and the second consists of one bit.

If $b_k, \ldots, b_l$ are strings, we let $\langle b_k, \ldots, b_l \rangle$ or, equivalently, $\langle b_j \rangle_{j=k}^{l}$ denote the $(l+1-k)$-element sequence whose ith element is b_{k+i-1}. An *input* is an element of $\{0,..,n-1\}\times\{0,..,n-1\}$, it denotes a value assignment for X and Y. For each input (x,y), a deterministic protocol determines a sequence $\langle b_1(x,y), \ldots, b_{m(x,y)}(x,y) \rangle$ of transmitted messages[4]. Since empty messages are allowed, we assume that $P_{\mathcal{X}}$ transmits the first message. Hence, all odd numbered messages are transmitted by $P_{\mathcal{X}}$ and all even messages, by $P_{\mathcal{Y}}$.

Initially, $P_{\mathcal{X}}$ knows only X, $P_{\mathcal{Y}}$ only knows Y, and their only information about each other's random variable derives from previosly transmitted messages. Therefore, for odd i's, $b_i(x,y)$ must be a function of x and $\langle b_j(x,y) \rangle_{j=1}^{i-1}$ only; the analogous statement holds for even i's.

A string $\alpha_1 \cdots \alpha_k$ of bits is called a *prefix* of a string $\beta_1 \cdots \beta_l$ of bits if $k \le l$ and $\alpha_i = \beta_i$ for $i = 1, \ldots, k$. It is called a *proper prefix* if, in addition, $k \ne l$. A set of strings is called *prefix free* if no string in the set is a prefix of another. In order for each person to know when the message he receives terminates, the set of possible messages $b_i(x,y)$ for any positive i must prefix free given all previous messages.

At the end of the communication, the $P_{\mathcal{X}}$ and $P_{\mathcal{Y}}$ decide on a *computed value* $v_\phi(X,Y)$, that they assume is the correct value of $f(X,Y)$ †. To enable $P_{\mathcal{X}}$ and $P_{\mathcal{Y}}$ to determine the computed value, $v_\phi(x,y)$ must be a function of x and $\langle b_j(x,y) \rangle_{j=1}^{m(x,y)}$, and also a function of y and $\langle b_j(x,y) \rangle_{j=1}^{m(x,y)}$.

With these requirements in mind, we now formally define deterministic protocols. To simplify notation and proofs, we extend the definition of $b_i(x,y)$ indefinitely and let $b_i(x,y) \stackrel{\text{def}}{=} \Lambda$ for all $i > m(x,y)$.

---

[3]Allowing empty messages simplifies notation and proofs. In addition, when there are more than two communicators (a case not considered here), empty messages eliminate the need for arbitrary transmission order.

[4]The messages depend of course on the protocol used, but, to shorten notation, we write $b_i(x,y)$ and $m(x,y)$ for $b_{\phi,i}(x,y)$ and $m_\phi(x,y)$.

†In general, each of $P_{\mathcal{X}}$ and $P_{\mathcal{Y}}$ could decide on a different computed value. However, any protocol ensuring that a given communicator (say $P_{\mathcal{X}}$) always knows $f(X,Y)$ can be augmented by one bit to yield a protocol ensuring that both $P_{\mathcal{X}}$ and $P_{\mathcal{Y}}$ always know $f(X,Y)$. (Note that requiring merely that at least one communicator know $f(X,Y)$ for all inputs devoids the problem of content).

A *deterministic protocol* ϕ *for* $\{0,..,n-1\} \times \{0,..,n-1\}$ can therefore be defined as a mapping that, with each input (x, y), associates:

1. a sequence $\langle b_1(x,y), b_2(x,y), \ldots, b_{m(x,y)}(x,y) \rangle$ of messages

2. a *computed value* $v_\phi(x, y)$

such that for all $x, x', y, y' \in \{0,..,n-1\}$ and for all $i \geq 1$:

Separate-transmissions property

for odd i, $\langle b_j(x,y) \rangle_{j=1}^{i-1} = \langle b_j(x,y') \rangle_{j=1}^{i-1}$ implies that $b_i(x,y) = b_i(x,y')$,
and,
for even i, $\langle b_j(x,y) \rangle_{j=1}^{i-1} = \langle b_j(x',y) \rangle_{j=1}^{i-1}$ implies that $b_i(x,y) = b_i(x',y)$.

Prefix-free-messages property

for odd i, $\langle b_j(x,y) \rangle_{j=1}^{i-1} = \langle b_j(x',y) \rangle_{j=1}^{i-1}$ implies that neither $b_i(x,y)$ nor $b_i(x',y)$ is a proper prefix of the other,
and,
for even i, $\langle b_j(x,y) \rangle_{j=1}^{i-1} = \langle b_j(x,y') \rangle_{j=1}^{i-1}$ implies that neither $b_i(x,y)$ nor $b_i(x,y')$ is a proper prefix of the other.

Separate-decisions property

$\langle b_j(x,y) \rangle_{j=1}^{\infty} = \langle b_j(x,y') \rangle_{j=1}^{\infty}$ implies that $v_\phi(x,y) = v_\phi(x,y')$,
and
$\langle b_j(x,y) \rangle_{j=1}^{\infty} = \langle b_j(x',y) \rangle_{j=1}^{\infty}$ implies that $v_\phi(x,y) = v_\phi(x',y)$.

Remarks:

1. Note that the first two properties must apply to all i's, even those larger than $m(x, y)$.

2. It can be shown that, although empty messages are allowed, the prefix-free-messages property does not increase the amount of communication required. Every protocol containing empty messages can be trivially modified to obtain a protocol that transmits the same number of bits but does not contain any empty message.

In the rest of this section we show that these requirements impose a structure on deterministic protocols, and then use this structure to prove some of their properties.

If $b_k, \ldots, b_l$ are strings, we let $[b_k \ldots b_l]$ or, equivalently, $[b_j]_{j=k}^{l}$ denote the string obtained by their concatenation. We say that the sequence $\langle b_1, \ldots, b_l \rangle$ *parses* the string b if $b = [b_1 \ldots b_l]$. In general, there are many ways to parse a string.

For every (x, y), we let $b_\phi(x, y) \stackrel{\text{def}}{=} [b_i(x, y)]_{i=1}^{m(x,y)}$, *the codeword associated with* (x, y), be the concatenation of all messages transmitted according to ϕ when (x, y) is the input. The next theorem shows that only one parsing of $b_\phi(x, y)$ complies with the separate-transmissions and the prefix-free-messages properties.

Theorem 1 If either $b_\phi(x, y)$ or $b_\phi(x', y')$ is a prefix of the other, then

$$\langle b_j(x, y) \rangle_{j=1}^{\infty} = \langle b_j(x', y) \rangle_{j=1}^{\infty} = \langle b_j(x, y') \rangle_{j=1}^{\infty} = \langle b_j(x', y') \rangle_{j=1}^{\infty}.$$

Proof: We show by induction on i that $\langle b_j(x, y) \rangle_{j=1}^{i} = \langle b_j(x', y) \rangle_{j=1}^{i} = \langle b_j(x, y') \rangle_{j=1}^{i} = \langle b_j(x', y') \rangle_{j=1}^{i}$. The induction hypothesis is clearly true for $i = 0$. Assume it is true for $i-1$; then, $[b_j(x, y)]_{j=1}^{i-1} = [b_j(x', y')]_{j=1}^{i-1}$. Hence, one of $[b_j(x, y)]_{j=i}^{\infty}$ and $[b_j(x', y')]_{j=i}^{\infty}$ is a prefix of the other, which implies that one of $b_i(x, y)$ and $b_i(x', y')$ is a prefix of the other. If i is odd, then, by the separate-transmissions property and the induction hypothesis, $b_i(x, y') = b_i(x, y)$. By the prefix-free-messages property, neither $b_i(x, y')$ nor $b_i(x', y')$ can be a proper prefix of the other and hence they must be the same. Using the same argument for (x', y), we get $b_i(x, y) = b_i(x', y) = b_i(x, y') = b_i(x', y')$. The symmetric argument will work for even i's, which completes the proof of the induction $\square$

Besides showing that there is a unique way to parse $b_\phi(x, y)$ into the individual messages, the theorem has three other important implications. They are used extensively later so we restate them as individual corollaries. To do so, we need the following definitions.

A *rectangle* is a set of the form $R_{\mathcal{X}} \times R_{\mathcal{Y}}$ where $R_{\mathcal{X}}, R_{\mathcal{Y}} \subseteq \{0,..,n-1\}$. If S is a subset of $\{0,..,n-1\} \times \{0,..,n-1\}$, we define its *closure* to be $\mathcal{C}\ell(S) \stackrel{\text{def}}{=} \{(x, y) :$ for some u and v in $\{0,..,n-1\}$, both (x, u) and (v, y) are in $S\}$. It can be shown that $\mathcal{C}\ell(S)$ is the intersection of all rectangles that contain S and is therefore a rectangle. The theorem implies that if $b_\phi(x, y) = b_\phi(x', y')$, then the "corner points" (x, y') and (x', y) have the same codeword.

Therefore, if a protocol assigns the same codeword to all inputs in a set S, it assigns the same codeword to all inputs in $\mathcal{C}\ell(S)$.

Let $R_\phi(x,y) \stackrel{\text{def}}{=} \{(u,v) : b_\phi(u,v) = b_\phi(x,y)\}$ be the set of inputs having the same codeword as (x,y). The first corollary of Theorem 1 states that $R_\phi(x,y)$ can assume only special forms.

Corollary 1 For all $(x,y) \in \{0,..,n-1\} \times \{0,..,n-1\}$, $R_\phi(x,y)$ is a rectangle.

Proof: By Theorem 1, $(u,v) \in \mathcal{C}\ell(R_\phi(x,y))$ implies that $b_\phi(u,v) = b_\phi(x,y)$, i.e., $(u,v) \in R_\phi(x,y)$. Thus, $\mathcal{C}\ell(R_\phi(x,y)) \subseteq R_\phi(x,y)$. But, $R_\phi(x,y)$ is always contained in $\mathcal{C}\ell(R_\phi(x,y))$. Hence, $R_\phi(x,y) = \mathcal{C}\ell(R_\phi(x,y))$ which is a rectangle $\square$

This corollary has an intuitive "near-proof". Consider the set S of inputs that can result in a given sequence of messages. By the separate-transmissions property, whenever $P_\mathcal{X}$ transmits a message, he partitions this set into ***horizontal slices*** (sets of the form $S \cap (A_i \times \{0,..,n-1\})$ where the A_i's partition S) – one slice corresponding to each message. (Note that this would not have been the case without the separate-transmissions property). Similarly, when $P_\mathcal{Y}$ transmits a message, he partitions S into *vertical slices* (sets of the form $S \cap (\{0,..,n-1\} \times A_\mathcal{Y})$). Since any horizontal (or vertical) slice of a rectangle is again a rectangle, it is clear, by induction, that $R_\phi(x,y)$ is a rectangle. The missing part in this "proof" is showing that different rectangles cannot correspond to the same concatenation of messages. This is the essence of Theorem 1.

We note that the argument of the last paragraph is used in [Yao 79]. It is correct there, because the model used is more restrictive: each message is one bit long. The argument shows that each message-sequence corresponds to a rectangle; in that model, each message is binary, and the number of messages is also the number of transmitted bits.

The second corollary states that the codewords associated with the rectangles form a prefix-free set. Let $\mathcal{R}_\phi \stackrel{\text{def}}{=} \{R_\phi(x,y) : (x,y) \in \{0,..,n-1\} \times \{0,..,n-1\}\}$ be the set of rectangles over which ϕ has a fixed codeword. Any input (x,y) belongs to the rectangle $R_\phi(x,y) \in \mathcal{R}_\phi$, and any two intersecting rectangles in $\mathcal{R}_\phi$ coincide. Therefore, $\mathcal{R}_\phi$ *partitions* $\{0,..,n-1\} \times \{0,..,n-1\}$ into $|\mathcal{R}_\phi|$ rectangles where $|S|$ denotes the cardinality of the set S. For each rectangle $R \in \mathcal{R}_\phi$, define $b_\phi(R)$ to be the unique codeword that ϕ assigns to all inputs in R. Then Theorem 1 implies:

Corollary 2 $\{b_\phi(R) : R \in \mathcal{R}_\phi\}$ is prefix free and its cardinality is $|\mathcal{R}_\phi|$ □

The third corollary combines the theorem with the separate-decisions property to show:

Corollary 3 For all $R \in \mathcal{R}_\phi$, the computed value, v_ϕ, is constant over R.

Proof: Let (x,y), (x',y') be two inputs in a rectangle $R \in \mathcal{R}_\phi$. From the theorem, $\langle b_j(x,y)\rangle_{j=1}^{\infty} = \langle b_j(x',y)\rangle_{j=1}^{\infty} = \langle b_j(x',y')\rangle_{j=1}^{\infty}$. By the separate-decisions property, $v_\phi(x,y) = v_\phi(x',y) = v_\phi(x',y')$ □

In summary, the three corollaries show that every deterministic protocol ϕ determines:

1. A partition $\mathcal{R}_\phi$ of $\{0,..,n-1\}\times\{0,..,n-1\}$ into rectangles. We let $R_\phi(x,y)$ denote the unique rectangle in $\mathcal{R}_\phi$ that contains (x,y).

2. A mapping $b_\phi : \mathcal{R}_\phi \to \{0,1\}^*$ where $b_\phi(R)$ for $R \in \mathcal{R}_\phi$ is the unique codeword that ϕ associates with all inputs in R. That is, for all $(x,y) \in \{0,..,n-1\}\times\{0,..,n-1\}$, $b_\phi(x,y) = b_\phi(R_\phi(x,y))$. The mapping b_ϕ must satisfy several requirements. Most germane to our purposes is that $\{b_\phi(R) : R \in \mathcal{R}_\phi\}$ is prefix free with cardinality $|\mathcal{R}_\phi|$. The length of $b_\phi(R)$ is denoted by $l_\phi(R)$.

3. A mapping $v_\phi : \mathcal{R}_\phi \to \{0,1\}$ where $v_\phi(R)$, for $R \in \mathcal{R}_\phi$, is the unique value that ϕ assigns to all the inputs (x,y) in R. That is, $v_\phi(x,y) = v_\phi(R_\phi(x,y))$ for all $(x,y) \in \{0,..,n-1\}\times\{0,..,n-1\}$.

In the rest of this chapter we apply these properties to prove lower bounds on the number of bits that must be transmitted in error-free computation of a function.

2.2 Worst-Case Complexity

So far, we have not addressed the relation between the computed value of a protocol and the function it is supposed to compute. To define the number of transmitted bits needed to compute a function, we need to do this first.

Denote the number of bits in a string b by $|b|$ and let $l_\phi(x,y) \stackrel{\text{def}}{=} |b_\phi(x,y)| = \sum_{i=1}^{m(x,y)} |b_i(x,y)|$ denote the total number of bits transmitted according to a protocol ϕ when $P_\mathcal{X}$ knows x and $P_\mathcal{Y}$ knows y. The *worst-case complexity of* ϕ is:

$$\hat{L}_\phi \stackrel{\text{def}}{=} \max\{l_\phi(x,y) : (x,y) \in \{0,..,n-1\}\times\{0,..,n-1\}\} .$$

If f is a function defined on $\{0,..,n-1\} \times \{0,..,n-1\}$ and ϕ is a protocol for $\{0,..,n-1\} \times \{0,..,n-1\}$, we say that ϕ is ***error-free*** for f if $v_\phi(x,y) = f(x,y)$ for all $(x,y) \in \{0,..,n-1\} \times \{0,..,n-1\}$. The ***worst-case communication complexity of*** f is then defined as:

$$\hat{C}_D(f,0) \stackrel{\text{def}}{=} \min\{\hat{L}_\phi(x,y) : \phi \text{ is an error-free, deterministic protocol for } f\} .$$

The superscript $\hat{}$ indicates worst-case complexity, the subscript $_D$ stands for deterministic complexity, and the 0 means that no errors are allowed (i.e., 0 probability of error). All three requirements will be relaxed later on.

According to Corollary 2, the set $\{b_\phi(R) : R \in \mathcal{R}_\phi\}$ is prefix free and of cardinality $|\mathcal{R}_\phi|$. This implies that $\hat{L}_\phi \geq \log |\mathcal{R}_\phi|$. Hence, a lower bound for the number of rectangles in $\mathcal{R}_\phi$ can serve to lower bound the worst-case complexity of ϕ.

If ϕ is error free, then, by Corollary 3, f must be constant over each rectangle in $\mathcal{R}_\phi$. We call a rectangle ***f-constant*** if f is constant over it, and call a partition of a set $S \subseteq \{0,..,n-1\} \times \{0,..,n-1\}$ into f-constant rectangles — an ***f-constant decomposition of*** S. Let $\mu_d(f)$, the ***decomposition number of*** f, denote the smallest number of rectangles in any f-constant decomposition of $\{0,..,n-1\} \times \{0,..,n-1\}$. From the above, $|\mathcal{R}_\phi| \geq \mu_d(f)$ for every error-free protocol ϕ. Hence, $\hat{C}_D(f,0) \geq \log \mu_d(f)$.

To lower bound $\mu_d(f)$, let λ_f^S, for a set $S \subseteq \{0,..,n-1\} \times \{0,..,n-1\}$, be the size of the largest subset T of S such that $C\ell(T)$ is f-constant:

$$\lambda_f^S \stackrel{\text{def}}{=} \max\{|T| : T \subseteq S, \text{ and } (x_1,y_1),(x_2,y_2) \in C\ell(T) \Rightarrow f(x_1,y_1) = f(x_2,y_2)\} .$$

Clearly, at least $|S|/\lambda_f^S$ rectangles in any f-constant decomposition of $\{0,..,n-1\} \times \{0,..,n-1\}$ are needed to contain S alone. Therefore:

Theorem 2 $\hat{C}_D(f,0) \geq \log\left(|S|/\lambda_f^S\right)$ for all $S \subseteq \{0,..,n-1\} \times \{0,..,n-1\}$ $\square$

Two special cases of this theorem have been often used in the literature. They are usually referred to as *fooling-set argument* and *largest f-constant rectangle argument.*

Fooling-Set Argument

Considering only sets S with $\lambda_f^S = 1$, we obtain:

Theorem 3 [LS 81] Let $S \subseteq \{0,..,n-1\} \times \{0,..,n-1\}$ be such that for all $(x,y),(x',y') \in S$, $C\ell\{(x,y),(x',y')\}$ is not f-constant. Then, $\hat{C}_D(f,0) \geq \log |S|$ $\square$

Sets S with $\lambda_f^S = 1$ are called ***fooling sets*** because if $(x_1, y_1), (x_2, y_2) \in S$ have the same codeword, then at least one of $P_{\mathcal{X}}$ and $P_{\mathcal{Y}}$ will be "fooled" by either (x_1, y_2) or (x_2, y_1) (see [Ul 84]).

Example 1 Let $S_{\mathcal{X}} = S_{\mathcal{Y}} = \{0,..,n-1\}$. The equality function was defined in the introduction as

$$equ(x,y) \stackrel{\text{def}}{=} \begin{cases} 1 & \text{if } x = y \\ 0 & \text{if } x \neq y \end{cases}$$

and the fooling-set argument was implicitly used: for $n \geq 2$, the set $\{(0,0),\ldots,(n-1,n-1)\} \cup \{(0,1)\}$, is clearly a fooling set, hence, $\hat{C}_D(equ, 0) \geq \lceil \log(n+1) \rceil$ □

On the other extreme, substituting $\{0,..,n-1\} \times \{0,..,n-1\}$ for S in Theorem 2, and letting $\lambda_f \stackrel{\text{def}}{=} \lambda_f^{\{0,..,n-1\} \times \{0,..,n-1\}}$ denote the size of the largest f-constant rectangle in $\{0,..,n-1\} \times \{0,..,n-1\}$, yields:

Largest f-Constant Rectangle Argument

Theorem 4 [Yao 79] $\hat{C}_D(f, 0) \geq \log \dfrac{n^2}{\lambda_f}$ □

Example 2 Let $n \stackrel{\text{def}}{=} 2^N$ for some integer N, and $S_{\mathcal{X}} = S_{\mathcal{Y}} = \{0,1\}^N$. For $x = x_1, \ldots, x_N$ and $y = y_1, \ldots, y_N$ in $\{0,1\}^N$, define the ***inner-product*** function by:

$$i.p.(x,y) \stackrel{\text{def}}{=} \bigoplus_{i=1}^{N} x_i y_i$$

where $\oplus$ denotes ***exclusive or***. It was shown in [PE 84] that the largest $i.p.$-constant rectangle is of size $n = 2^N$. Using Theorem 4, this implies that $\hat{C}_D(i.p., 0) \geq N = \log n$. We now provide a simpler derivation for the size of the largest $i.p.$-constant rectangle. First, modify the inner-product function to have the range ± 1 by letting:

$$i.p.^\star(x,y) \stackrel{\text{def}}{=} \begin{cases} 1 & \text{if } i.p.(x,y) = 1 \\ -1 & \text{if } i.p.(x,y) = 0. \end{cases}$$

We will see in Lemma 6 that for every rectangle $R \subseteq \{0,..,n-1\} \times \{0,..,n-1\}$,

$$\Big| \sum_{\{(x,y) \in R\}} i.p.^\star(x,y) \Big| \leq \sqrt{n|R|}\ .$$

Therefore any *i.p.*-constant rectangle R must satisfy:

$$|R| = |\sum_{\{(x,y)\in R\}} i.p.^*(x,y)| \le \sqrt{n|R|} \,.$$

Hence, $|R| \le n$. The *i.p.*-constant rectangle $\{0\}\times\{0,..,n{-}1\}$ proves that the bound is tight □

Let $\{0,1\}^{\{0,..,n-1\}\times\{0,..,n-1\}}$ denote the set of boolean functions defined on $\{0,..,n-1\}\times\{0,..,n-1\}$. The largest f-constant rectangle argument was used by A. Yao to show that

Corollary 4 [Yao 79] Most functions in $\{0,1\}^{\{0,..,n-1\}\times\{0,..,n-1\}}$ have

$$\hat{C}_D(f,0) \ge \log n - 2 \,.$$

Proof: Fix a rectangle of area $\ge A$ in $\{0,..,n-1\}\times\{0,..,n-1\}$. The number of functions constant over this rectangle is $\le 2\cdot 2^{n^2-A}$. The number of functions constant over *some* rectangle of area $\ge A$ is therefore at most $2\cdot 2^{n^2-A}\cdot 2^n\cdot 2^n$. Thus the fraction of functions with $\lambda_f \ge A$ is at most $2\cdot 2^{2n-A}$. Taking $A = 4n$, a fraction of at least $1-2\cdot 2^{-2n}$ of the functions in $\{0,1\}^{\{0,..,n-1\}\times\{0,..,n-1\}}$ have $\hat{C}_D(f,0) \ge \log\frac{n^2}{4n} = \log n - 2$ □

Which of the "fooling set" and the "largest f-constant rectangle" is stronger? Neither. Continuing along the lines of [AUY 83], the equality function has a constant rectangle of size $n^2/4$ (lower bound of 2 bits) but still has a fooling set of size n (lower bound of $\log n$ bits). On the other hand, a simple argument can show that for a fraction of $1 - \frac{1}{n^{\log\log n}}$ of the functions, the largest rectangle is of size $4n$ (lower bound $\log n - 2$) while the largest fooling set is of size $\le 4\log n$ (lower bound of $\log\log n$). In fact, in part of the proof of the next theorem, the bounds derived using both methods are weak, but considering sets with $\lambda_f^S = 2$ and using Theorem 2, yields tight bounds (see [OE 86]).

Both Theorem 2 and the two arguments, lower bound $\log\mu_d(f)$ and use the inequality $\hat{C}_D(f,0) \ge \log\mu_d(f)$. However, the discussion following Corollary 1 implies that with each transmitted message, $P_{\mathcal{X}}$ and $P_{\mathcal{Y}}$ shift from computing $f|_R$ (the restriction of f to R) for some rectangle R to computing one of $f|_{R_1},\ldots,f|_{R_J}$ where the R_j's are either vertical or horizontal slices of R. Clearly, not all rectangles R can be partitioned into slices $R_1, \ldots, R_J$ such that $\mu_d(f|_R) = \mu_d(f|_{R_1}) + \ldots + \mu_d(f|_{R_J})$. Therefore, the existence of an f-constant decomposition of $\{0,..,n-1\} \times \{0,..,n-1\}$ into $\mu_d(f)$ rectangles, does not imply that $\hat{C}_D(f,0) = \lceil \log \mu_d(f) \rceil$.

It is therefore of interest to find the largest possible discrepancy between $\log \mu_d(f)$ and $\hat{C}_D(f,0)$. The first result in this direction, proved in [Yao 79], showed that $\hat{C}_D(f,0) \leq O\left(\sqrt{\mu_d(f)} \log \mu_d(f)\right)$ and was later improved:

Theorem 5 [AUY 83] $\quad \hat{C}_D(f,0) \leq O(\log^2 \mu_d(f))$.

Proof: Let $\mathcal{R}$ be an f-constant decomposition of $\{0,..,n-1\} \times \{0,..,n-1\}$. Define two graphs $G_\mathcal{X}$ and $G_\mathcal{Y}$ both having the set $\mathcal{R}$ of rectangles as nodes. There is an edge in $G_\mathcal{X}$ between two rectangles $R = R_\mathcal{X} \times R_\mathcal{Y}$ and $Q = Q_\mathcal{X} \times Q_\mathcal{Y}$ if R and Q have a row in common, i.e., if $R_\mathcal{X} \cap Q_\mathcal{X} \neq \emptyset$. Similarly, there is an edge in $G_\mathcal{Y}$ if $R_\mathcal{Y} \cap Q_\mathcal{Y} \neq \emptyset$.

For every subset $\mathcal{V} \subseteq \mathcal{R}$, let $G_\mathcal{X}(\mathcal{V})$ (respectively, $G_\mathcal{Y}(\mathcal{V})$) be the restriction of $G_\mathcal{X}$ (resp. $G_\mathcal{Y}$) to $\mathcal{V}$ and let $deg_{G_\mathcal{X}(\mathcal{V})}(R)$ be the degree of the node R in the graph $G_\mathcal{X}(\mathcal{V})$ (similarly for $deg_{G_\mathcal{Y}(\mathcal{V})}(R)$).

Since no two rectangles in $\mathcal{V}$ can have both a row and a column in common, we have $deg_{G_\mathcal{X}(\mathcal{V})}(R) + deg_{G_\mathcal{Y}(\mathcal{V})}(R) \leq |\mathcal{V}|$ for all rectangles $R \in \mathcal{V}$. The protocol proceeds as follows: starting with $\mathcal{V} = \mathcal{R}$, $P_\mathcal{X}$ and $P_\mathcal{Y}$ repeat the following iteration until $\mathcal{V}$ consists of exactly one rectangle.

$P_\mathcal{X}$ looks for a rectangle $R = R_\mathcal{X} \times R_\mathcal{Y} \in \mathcal{V}$ such that $X \in R_\mathcal{X}$ and $deg_{G_\mathcal{X}(\mathcal{V})}(R) \leq |\mathcal{V}|/2$. If he finds such a rectangle, he transmits 1 followed by the index of this rectangle. $P_\mathcal{Y}$ then responds by sending 1 if $Y \in R_\mathcal{Y}$ and 0 otherwise. They then update $\mathcal{V}$ to consist of the rectangle R if $P_\mathcal{Y}$'s bit was 1, and of the rectangles that share a common row with R if $P_\mathcal{Y}$'s transmitted 0.

If $P_\mathcal{X}$ cannot find such a rectangle, he transmits 0 and, then, $P_\mathcal{Y}$ transmits the index of a rectangle $R = R_\mathcal{X} \times R_\mathcal{Y} \in \mathcal{V}$ such that $Y \in R_\mathcal{Y}$ and $deg_{G_\mathcal{Y}(\mathcal{V})}(R) \leq |\mathcal{V}|/2$. (Such a rectangle must exist because, from the above, the rectangle $R(X,Y)$ that contains (X,Y) satisfies: $deg_{G_\mathcal{X}(\mathcal{V})}(R(X,Y)) + deg_{G_\mathcal{Y}(\mathcal{V})}(R(X,Y)) \leq |\mathcal{V}|$.) $P_\mathcal{X}$ then responds with one bit telling $P_\mathcal{Y}$ whether $X \in R_\mathcal{X}$. And they update $\mathcal{V}$ as before.

Clearly, $R(X,Y) \in \mathcal{V}$ at all times. Also, the number of rectangles in $\mathcal{V}$ decreases by at least a factor of 2 at each iteration, hence the number of iterations is at most

$\log|\mathcal{R}|$. The number of bits transmitted in each iteration is at most $2+\log|\mathcal{R}|$. Therefore, $\hat{L}_\phi \leq \log^2|\mathcal{R}| + 2\log|\mathcal{R}|$. All they need to do to achieve the claimed bound is pick an f-constant decomposition $\mathcal{R}$ of size $\mu_d(f)$ □

In fact, [AUY 83] proved a stronger result that holds for rectangles that can overlap each other. The interesting problem of whether the theorem can be further improved is still open (see subsection 5.1).

A special case does, however, have a tight solution. An *interval mod* n is a subset of $\{0,..,n-1\}$ of the form $\{(I+j)_{\text{mod } n} : j = 0,\dots,J\}$. Earlier, a rectangle was defined as a product set of the form $R_\mathcal{X} \times R_\mathcal{Y}$ where $R_\mathcal{X}, R_\mathcal{Y} \subseteq \{0,..,n-1\}$. A rectangle is *planar* if $R_\mathcal{X}$ and $R_\mathcal{Y}$ are intervals mod n. If $\{0,..,n-1\}\times\{0,..,n-1\}$ is drawn on a torus in the obvious way, any two points in a planar rectangle can be connected by a line that is contained in that rectangle. We call a decomposition of $\{0,..,n-1\}\times\{0,..,n-1\}$ into rectangles *planar* if all the rectangles in the decomposition are planar.

Let $\mu_p(f)$ be the minimal number of rectangles in any f-constant planar decomposition of $\{0,..,n-1\}\times\{0,..,n-1\}$. In general, $\mu_d(f) \leq \mu_p(f) \leq 2^{\mu_d(f)}$. It was shown in [Yao 79] that $\hat{C}_D(f,0) \leq \frac{2\log\mu_p(f)}{\log(4/3)} + 6$. We now provide a simple proof (done with K. Pang) that shows:

Lemma 1 $\hat{C}_D(f,0) \leq \lceil \log\mu_p(f)\rceil + 1$.

Proof: Denote each nonempty interval $\{(I+j)_{\text{mod } n} : j = 0,\dots,J\}$ by $[I, I+J]$ (to ensure uniqueness, the interval $\{0,\dots,n-1\}$ is always denoted by $[0, n-1]$).

$P_\mathcal{X}$ and $P_\mathcal{Y}$ agree a priori on a decomposition of $\{0,..,n-1\}\times\{0,..,n-1\}$ into $\mu_p(f)$ f-constant planar rectangles: $[\alpha_j,\beta_j]\times[\gamma_j,\delta_j]$ for $j = 1,\dots,\mu_p(f)$. The communication proceeds as follows:

1) $P_\mathcal{X}$ transmits $g(X) = \begin{cases} |\{\alpha_j : \alpha_j \leq X\}| - 1 & \text{if } \exists \alpha_j \leq X \\ |\{\alpha_j : j = 1,\dots,\mu_p(f)\}| & \text{if } \forall \alpha_j > X. \end{cases}$
2) $P_\mathcal{Y}$ picks the smallest X' such that $g(X') = g(X)$ and transmits $f(X',Y)$.
3) They take the bit transmitted by $P_\mathcal{Y}$ to be the computed value.

The number of bits transmitted is at most $\lceil\log\mu_p(f)\rceil + 1$. Correctness is guaranteed because for all x_1, x_2, and y, $g(x_1) = g(x_2)$ implies that $f(x_1,y) = f(x_2,y)$. Note that the lemma holds even if we allow row and column permutations in the definition of planar decompositions □

Yet another approach to lower bounds was introduced in [MS 82]. The *rank* of a matrix is the largest number of linearly independent rows in the matrix. If A is a matrix, denote its (x,y)th element by $A(x,y)$. To be consistent with definitions made so far, we assume that the first row and the first column of A are numbered 0. The *function matrix* M_f of a function $f \in \{0,1\}^{\{0,..,n-1\}\times\{0,..,n-1\}}$ is the $n \times n$ binary matrix defined by $M_f(x,y) \stackrel{\text{def}}{=} f(x,y)$ for $x, y \in \{0,..,n-1\}$.

It was shown in [MS 82] that, for the restricted model where $P_{\mathcal{X}}$ and $P_{\mathcal{Y}}$ are constrained to transmitting one bit at a time, the worst-case complexity is at least log rank(M_f). The proof involved the use of nondeterministic computations and was significantly simplified by M. Paterson. [AUY 83] noted that $\mu_d(f) \geq \text{rank}(M_f)$ thereby further simplifying the proof:

Theorem 6 [MS 82,AUY 83] $\hat{C}_D(f,0) \geq \log \mu_d(f) \geq \log\ \text{rank}(M_f)$.

Proof: Let $R_1, \ldots, R_{\mu_d(f)}$ be an f-constant decomposition of $\{0,..,n-1\} \times \{0,..,n-1\}$. Then,

$$f(x,y) = \sum_{k=1}^{\mu_d(f)} f(R_k)\chi_{R_k}(x,y) \tag{1}$$

where $f(R_k)$ is the unique value that f assumes over R_k and

$$\chi_{R_k}(x,y) \stackrel{\text{def}}{=} \begin{cases} 1 & \text{if } (x,y) \in R_k \\ 0 & \text{if } (x,y) \notin R_k \end{cases}$$

is the characteristic function of R_k. Equation (1) can be rewritten in matrix form as

$$M_f = \sum_{k=1}^{\mu_d(f)} f(R_k) M_{\chi_{R_k}}.$$

Using the subadditivity of ranks, and the inequality rank$(M_{\chi_{R_k}}) \leq 1$, we obtain:

$$\text{rank}(M_f) = \text{rank}\left(\sum_{k=1}^{\mu_d(f)} f(R_k)M_{\chi_{R_k}}\right) \leq \sum_{k=1}^{\mu_d(f)} \text{rank}\left(f(R_k)M_{\chi_{R_k}}\right) \leq \mu_d(f) \qquad \square$$

Let $\mathbf{1}_f \stackrel{\text{def}}{=} \{(x,y) : f(x,y) = 1\}$ denote the set of inputs for which f is 1. It is clear from the proof that rank(M_f) is smaller than the size of any partition of $\mathbf{1}_f$ into rectangles. Letting

$$\overline{f}(x,y) \stackrel{\text{def}}{=} \begin{cases} 1 & \text{if } f(x,y) = 0 \\ 0 & \text{if } f(x,y) = 1 \end{cases}$$

be the complement of f, it is equally clear that $\text{rank}(M_{\bar{f}})$ is smaller than the size of any rectangular partition of $\mathbf{0}_f \stackrel{\text{def}}{=} \{(x,y) : f(x,y) = 0\}$. Therefore,

$$\mu_d \geq \text{rank}(M_f) + \text{rank}(M_{\bar{f}}) \ . \tag{2}$$

In the next two examples, we use this bound to obtain the exact complexity of the inner product and the equality functions.

Example 3 Let $i.p.$ be as in Example 2. [AUY 83] showed that the 2^N by 2^N matrix $M_{i.p.}$ has rank N over $GF(2)$ which implies that $\hat{C}_D(i.p., 0) \geq \log N$. Over the reals, $\text{rank}(M_{i.p.}) = 2^N - 1$ which yields the bound of Example 2: $\hat{C}_D(i.p., 0) \geq N$. The exact bound is obtained by considering $i.p.^\star$. The discussion preceding Lemma 6 shows that $M_{i.p.^\star}$ is a Hadamard matrix over the reals (see there for definitions) and therefore $\text{rank}(M_{i.p.^\star}) = \text{rank}(M_{\overline{i.p.^\star}}) = n$. Using Equation (2), $\hat{C}_D(i.p., 0) = \hat{C}_D(i.p.^\star, 0) = N + 1$ □

Example 4 In example 1, we showed that n rectangles are needed to partition the set $\mathbf{1}_{equ} = \{(0,0), \ldots, (n-1, n-1)\}$. As you have probably discovered if you tried to solve the first puzzle posed in the introduction, it is much harder to find a similar argument showing that n rectangles are needed to partition $\mathbf{0}_{equ} = \{(i,j) \in \{0,..,n-1\} \times \{0,..,n-1\} : i \neq j\}$. Yet, over the reals, $\text{rank}(M_{\overline{equ}}) = n$. Hence $\mu_d(equ) = 2n$, and

$$\hat{C}_D(equ, 0) = \lceil \log n \rceil + 1 \ .$$

A similar argument can show that each rectangle in a rectangular partition of $\mathbf{0}_{equ}$ into n rectangles is either a "row" or a "column": replacing a rectangle of 1's in $M_{\overline{equ}}$ by a rectangle of 0's does not reduce the rank of the matrix, unless the rectangle is a "row" or a "column" □

Let $\mathbf{1}_f \stackrel{\text{def}}{=} \{(x,y) : f(x,y) = 1\}$ be the set of inputs for which f is 1, and $\mathbf{0}_f$ — the set of inputs for which f is 0. For every integer $0 \leq s \leq n^2$, define

$$\mathcal{F}_s \stackrel{\text{def}}{=} \{f \in \{0,1\}^{\{0,..,n-1\} \times \{0,..,n-1\}} : |\mathbf{1}_f| = s\}$$

(the set of boolean functions defined over $\{0,..,n-1\} \times \{0,..,n-1\}$ that have s ones). The number of functions in $\mathcal{F}_s$ is $\binom{n^2}{s}$, hence the vast majority of functions belong to $\mathcal{F}_s$ for

$s \approx n^2/2$. Consequently, Corollary 4 does not imply that most *sparse functions* ($s \ll n^2/2$) have high complexity. The following refinement says that even functions with very few ones are likely to have high complexity.

Theorem 7 [OE 86] For any $n \leq s \leq n^2/2$, most functions in $\mathcal{F}_s$ have

$$\hat{C}_D(f,0) \geq \log n - 4 \quad \Box$$

This Theorem is of interest because, as we shall see in the next subsection, most functions have average complexity also about $\log n$, but most sparse functions have very low average complexity. The contrast between the two results displays a crucial difference between worst-case and average communication complexity.

2.3 Average Complexity

The worst-case complexity of a protocol ϕ was defined in the last subsection as the number of bits transmitted according to ϕ for the "worst" input. The *average complexity of a protocol* ϕ is defined as the expected number of transmitted bits when X, Y are uniformly distributed over $\{0,..,n-1\} \times \{0,..,n-1\}$:

$$\bar{L}_\phi \stackrel{\text{def}}{=} \frac{1}{n^2} \sum_{(x,y)\in\{0,..,n-1\}\times\{0,..,n-1\}} l_\phi(x,y) \; .$$

The *average, deterministic communication complexity of* f is defined as:

$$\bar{C}_D(f,0) \stackrel{\text{def}}{=} \min\{\bar{L}_\phi(x,y) : \phi \text{ is an error-free, deterministic protocol for } f\} \; .$$

The subscript D and the 0 have the same meaning as before. The bar above the C indicates average complexity.

The fooling-set argument of the last section is "local": a small set of inputs can yield a large lower bound. It cannot therefore lower bound average communication complexity. Indeed, it was used in Example 1 to show that $\hat{C}_D(equ,0) \geq \log n$, while protocol ϕ_2 of the introduction implies that $\bar{C}_D(equ,0) \leq 4$. The largest f-constant rectangle argument, on the other hand, is more global and, with the aid of a simple information-theoretic result, can be used to lower bound average complexity.

A probability distribution is a set of non-negative reals that add to one. To allow multiple equal probabilities, we regard probability distributions as multisets, but for lack of alternative notation we denote them, as sequences, by angled brackets. The *entropy* of a probability distribution $\langle p(u) : u \in \mathcal{U} \rangle$ is defined as

$$H \langle p(u) : u \in \mathcal{U} \rangle \stackrel{\text{def}}{=} \sum_{u \in \mathcal{U}} p(u) \log \frac{1}{p(u)}.$$

The next information-theoretic lemma says that the number of bits needed to describe the outcome of a random variable is at least the entropy of the underlying probability distribution.

Lemma 2 [Sh 48] Let $\langle p_j : j = 1, \ldots, J \rangle$ be a probability distribution and $\langle l_j : j = 1, \ldots, J \rangle$ be the lengths of codewords in a prefix-free code[3]. Then,

$$\sum_{j=1}^{J} p_j l_j \geq H \langle p_j : j = 1, \ldots, J \rangle \quad \square$$

The lemma can be readily used to relate the sizes of the rectangle in $\mathcal{R}_\phi$ to $\bar{L}_\phi$.

Theorem 8 If ϕ is a deterministic protocol, then $\bar{L}_\phi \geq H \langle |R|/n^2 : R \in \mathcal{R}_\phi \rangle$.

Proof:

$$\begin{aligned} \bar{L}_\phi &\stackrel{\text{def}}{=} \sum_{(x,y) \in \{0,..,n-1\} \times \{0,..,n-1\}} \frac{1}{n^2} l_\phi(x,y) \\ &= \sum_{R \in \mathcal{R}_\phi} \sum_{(x,y) \in R} \frac{1}{n^2} l_\phi(x,y) \\ &= \sum_{R \in \mathcal{R}_\phi} \sum_{(x,y) \in R} \frac{1}{n^2} l_\phi(R) \\ &= \sum_{R \in \mathcal{R}_\phi} l_\phi(R) \frac{|R|}{n^2}. \end{aligned}$$

But, from Lemma 2, $\langle l_\phi(R) : R \in \mathcal{R}_\phi \rangle$ are lengths of codewords in a prefix-free code. Hence,

$$\sum_{R \in \mathcal{R}_\phi} l_\phi(R) \frac{|R|}{n^2} \geq H \left\langle |R|/n^2 : R \in \mathcal{R}_\phi \right\rangle \quad \square$$

[3]A set of strings is traditionally called *a codebook* or a *code*. Its elements are called codewords.

This theorem can be used to show that the largest f-constant rectangle argument of Corollary 4 holds even for average complexity, thus simplifying the proof provided in [JK 84].

Corollary 5 [JK 84]† $\quad \bar{C}_D(f,0) \geq \log \frac{n^2}{\lambda_f}$.

Proof: For every deterministic protocol ϕ and every rectangle $R \in \mathcal{R}_\phi$, $\quad |R|/n^2 \leq \lambda_f/n^2$. Therefore, $H\langle |R|/n^2 : R \in \mathcal{R}_\phi \rangle \geq \log \frac{n^2}{\lambda_f}$ □

Example 5 Consider the inner-product function of Examples 2 and 3 once again. Both examples proved that $\hat{C}_D(i.p.,0) \geq \log n$. By the corollary, the argument of Example 2 also proves that $\bar{C}_D(i.p.,0) \geq \log n$ □

As in Corollary 4, the last corollary implies that most functions have $\bar{C}_D(f,0) \geq \log n - 2$. However, unlike worst-case complexity (Theorem 7), this is not the case for most functions in $\mathcal{F}_s$ when $s \ll n^2/2$. In fact, it can be shown that

Theorem 9 [OE 86] For all $n \leq s \leq n^2/2$:

1) *all* functions f in $\mathcal{F}_s$ have $\bar{C}_D(f,0) \leq \log \frac{s}{n} + 8.3 \log\log n + c$,

2) most functions f in $\mathcal{F}_s$ have $\bar{C}_D(f,0) \geq \log \frac{s}{n} - 2\log\log \frac{s}{n}$ □

Hence, typically, average communication complexity decreases for sparse functions while worst-case complexity remains about $\log n$.

3 Randomized Protocols

Randomized protocols are much like deterministic ones, except that $P_\mathcal{X}$ and $P_\mathcal{Y}$ may use coin flips to determine their transmissions. Whereas deterministic protocols require that $P_\mathcal{X}$ base his transmissions and his computed value only on X and preceding transmissions (similarly for $P_\mathcal{Y}$), randomized protocols require that $P_\mathcal{X}$ base the bias of the coin flips (which determine his transmissions and his computed value) only on X and preceding transmissions (similarly for $P_\mathcal{Y}$). Still, the set of all messages that have positive probability at any time must be prefix free. And, the computed values must always agree (this requires a simple rejustification).

---

†They showed that $\bar{C}_D(f,0) \geq \Omega(\log \frac{n^2}{\lambda_f})$.

We denote randomized protocols by Φ (as opposed to ϕ for deterministic protocols). The number of bits transmitted when $P_{\mathcal{X}}$ knows x and $P_{\mathcal{Y}}$ knows y is now a random variable denoted by $L_\Phi(x,y)$; its expected value is denoted by $\bar{L}_\Phi(x,y)$.

The definitions of average and worst-case complexities of protocols can now be extended to randomized protocols. The *average complexity* $\bar{L}_\Phi$ *of* Φ is:

$$\bar{L}_\Phi \stackrel{\text{def}}{=} \frac{1}{n^2} \sum_{(x,y)\in\{0,..,n-1\}\times\{0,..,n-1\}} \bar{L}_\Phi(x,y).$$

The *worst-case complexity* $\hat{L}_\Phi$ *of* Φ is:

$$\hat{L}_\Phi \stackrel{\text{def}}{=} \max\{\bar{L}_\Phi(x,y) : (x,y) \in \{0,..,n-1\}\times\{0,..,n-1\}\}.$$

Similarly, the *computed value*, $V_\Phi(x,y)$, which $P_{\mathcal{X}}$ and $P_{\mathcal{Y}}$ assume is the value of $f(x,y)$ is now a 0-1 random variable and we let $\bar{E}_\Phi(x,y) \stackrel{\text{def}}{=} p(V_\Phi(x,y) \neq f(x,y))$ denote the probability that the computed value determined by Φ is wrong for (x,y). (The definition extends to deterministic protocols by letting $p(v_\phi(x,y) \neq f(x,y)) \stackrel{\text{def}}{=} \chi_{\{(x,y):v_\phi(x,y)\neq f(x,y)\}}(x,y)$ where $\chi_S(x,y)$ is the characteristic (or indicator) function of S that is 1 inside S and 0 elsewhere.) The *average error incurred by* Φ *in computing* f is defined as:

$$\bar{E}_\Phi(f) \stackrel{\text{def}}{=} \frac{1}{n^2} \sum_{(x,y)\in\{0,..,n-1\}\times\{0,..,n-1\}} \bar{E}_\Phi(x,y).$$

The *worst-case error incurred by* Φ *in computing* f is defined as:

$$\hat{E}_\Phi(f) \stackrel{\text{def}}{=} \max\{\bar{E}_\Phi(x,y) : (x,y) \in \{0,..,n-1\}\times\{0,..,n-1\}\}.$$

Remark: The maximum in $\hat{L}_\Phi$ and $\hat{E}_\Phi(f)$ is taken over the inputs. For each input, we still average over the "coin flips".

Using these quantities, we define two more communication-complexity measures. The *worst-case, randomized communication complexity of a function* f *with* ϵ *error*:

$$\hat{C}_R(f,\hat{\epsilon}) \stackrel{\text{def}}{=} \min\{\hat{L}_\Phi : \hat{E}_\Phi(f) \leq \epsilon\},$$

and the *average, randomized complexity of* f *with* ϵ *error*:

$$\bar{C}_R(f,\hat{\epsilon}) \stackrel{\text{def}}{=} \min\{\bar{L}_\Phi : \hat{E}_\Phi(f) \leq \epsilon\}.$$

Again, $\hat{C}$ denotes worst-case complexity while $\bar{C}$ stands for average complexity and the subscript $_R$ indicates that randomized protocols are allowed. The hat in $\hat{\epsilon}$ means that ϵ worst-case error is permitted. We only consider worst-case errors for reasons mentioned in the introduction.

The first interesting result concerning randomized protocols states that randomization and errors can help. It was described in the introduction and is presented here in more detail. Recall that the worst-case error-free complexity of the equality function is $\log n$ bits. [RY 79] (see [Yao 79]) used the prime numbers algorithm of [Fr 77] to show that $\hat{C}_R(equ, \hat{\epsilon}) \le O(\log\log n)$. The following theorem is more specific.

Theorem 10 [Yao 79] $\quad \hat{C}_R(equ, \epsilon) \le 2(\log\ln n + \log\frac{1}{\epsilon} + 1)$.

Proof: Let $\sigma \stackrel{\text{def}}{=} \sqrt{2}\frac{\ln n}{\epsilon}$. If $n \le 60,000$, then $2(\log\ln n + \log\frac{1}{\epsilon}+1) > 1+(1-2\epsilon)(\log n+1)$ which is the number of bits required by the simple minded protocol. On the other hand, $n > 60{,}000$ implies $\sigma > 31.1$ so the number of primes between σ and 2σ is at least $\frac{\sigma}{\sqrt{2}\ln\sigma}$ and at most $\frac{\sigma}{\ln\sigma}$. The number of primes between σ and 2σ that divide any positive integer $\le n$ is always at most $\log_\sigma n$.

For brevity, let $(x)_\alpha$ denote $x \bmod \alpha$. The following protocol achieves the claimed bound.

1) $P_{\mathcal{Y}}$ picks at random a prime α such that $\sigma < \alpha < 2\sigma$. He transmits α and $(X)_\alpha$.
2) $P_{\mathcal{X}}$ transmits 1 if $(X)_\alpha = (Y)_\alpha$ and 0 otherwise.
3) They take the bit transmitted by $P_{\mathcal{X}}$ to be the computed value.

For each input (x, y) such that $x = y$, the computed value is always $1 = equ(x, y)$. Hence $P(V_{\Phi_1}(x,y) = 0 | (x,y) \in \mathbf{1}_f) = 0$. On the other hand,

$$\begin{aligned} P\Big(V_{\Phi_1}(x,y) = 1 | x \ne y\Big) &= P\Big((x)_\alpha = (y)_\alpha | x \ne y\Big) \\ &\le \log_\sigma n \Big/ \tfrac{\sigma}{\sqrt{2}\ln\sigma} \\ &\le \sqrt{2}m\tfrac{\ln n}{\sigma} = \epsilon\,. \end{aligned}$$

The number of bits transmitted is upper bounded as follows:

Transmitting α : $\log\frac{\sigma}{\ln\sigma} \le \log\ln n + \log\frac{1}{\epsilon} + .5 - \log\ln 31.1$ bits.

Transmitting y mod α: $\overline{\log\alpha} \le \log\overline{\alpha} \le \log 1.7\sigma \le \log\ln n + \log\frac{1}{\epsilon} + .5 + .77$ bits.

Rounding-off the above: at most 1 bit.

Transmitting whether $(x)_\alpha = (y)_\alpha$: 1 bit.

Hence the total number of exchanged bits is at most $2(\log\ln n + \log\frac{1}{\epsilon} + 1)$ bits $\square$

The complexity reduction for this function stems both from using randomization and allowing errors. Neither randomization nor errors alone can help reduce the complexity from $\log n$ (see Example 7). The next example shows that, for some functions, randomization can reduce the worst-case complexity even in error-free computation.

Example 6 [MS 82] Let $M \geq 0$, and $x^1,\ldots,x^M,y^1,\ldots,y^M \in \{0,1\}^M$ (we index the sequences with superscripts because subscripts were used earlier to denote a bit in a sequence). If $n = 2^{M^2}$, then each $(x^1,\ldots,x^M)$ can be identified with an integer in $\{0,..,n-1\}$. The *component equality* function is defined as:

$$c.e.\big((x^1,\ldots,x^M),(y^1,\ldots,y^M)\big) \stackrel{\text{def}}{=} \begin{cases} 1 & \text{if } x^i = y^i \text{ for some } 1 \leq i \leq M \\ 0 & \text{otherwise.} \end{cases}$$

[MS 82] used the rank of $M_{c.e.}$ to show that $\hat{C}_D(c.e.,0) \geq M^2$ and proved that $\hat{C}_R(c.e.,0) \leq O(M\log^2 M)$. The following protocol shows that $\hat{C}_R(c.e.,0) \leq 5M\log M + c$.

1) For $i = 1,\ldots,M$:

   a) $P_{\mathcal{X}}$ and $P_{\mathcal{Y}}$ compute $equ(X^i,Y^i)$ using the protocol of Theorem 10 with $\epsilon = \frac{\log M}{M}$.

   b) If they conclude that $X^i \neq Y^i$, they move to step (a) for the next iteration. Otherwise, they exchange X^i and Y^i and check whether they are indeed equal. They stop if $X^i = Y^i$ and move to step (a) for the next two sequences otherwise.

2) The computed value is 1 if for some i, $P_{\mathcal{X}}$ and $P_{\mathcal{Y}}$ exchanged X^i and Y^i and found them equal. It is 0 otherwise.

Clearly, communication stops precisely after the first i such that $x^i = y^i$. The number of bits transmitted in this iteration is $M + 1 + 2(\ln M + \log\frac{M}{\log M} + 1)$. The expected number of bits transmitted in each of the previous iterations is at most $M \cdot 2(\ln M + \log\frac{M}{\log M} + 1)$ $\square$

To prove lower bounds on randomized protocols, we use the following simple relationship between randomized and deterministic protocols. Let Φ be a randomized protocol. Each transmission is the outcome of a random experiment whose probability distribution is determined by Φ according to the integer known to the transmitter and previous transmissions.

Conceptually, each communicator can perform all the random experiments (corresponding to all possible integers he might have and all possible transmissions) prior to the commencement of the communication. Then, given his value and previous transmissions, he consults the appropriate experiment and discards the rest.

These two huge random experiments, each performed by one communicator, are independent of the current input. Every combined outcome of these two random experiments induces a deterministic protocol. Let $p(\phi)$ be the total probability of the combined outcomes that induce the deterministic protocol ϕ. As before, each ϕ determines $l_\phi(x,y)$ and $v_\phi(x,y)$. The expected communication length for (x,y), denoted earlier in this subsection as $\bar{L}_\Phi(x,y)$, and the probability of error $\bar{E}_\Phi(x,y)$ can now be computed as:

$$\begin{aligned} \bar{L}_\Phi(x,y) &= \sum_\phi p(\phi) l_\phi(x,y) \,, \\ \bar{E}_\Phi(x,y) &= \sum_{\{\phi : v_\phi(x,y) \neq f(x,y)\}} p(\phi). \end{aligned} \tag{3}$$

These equations clearly imply that:

$$\begin{aligned} \bar{L}_\Phi &= \sum_\phi p(\phi) \bar{L}_\phi \,, \\ \bar{E}_\Phi(f) &= \sum_\phi p(\phi) \bar{E}_\phi(f). \end{aligned} \tag{4}$$

Before proceeding further, we note that there are two possible models.

1) *Shared random sources.* In which the communicators have access to each other's random variable.

2) *Separate random sources.* In which the communicators do not have access to each other's random variable.

Clearly, $\hat{C}_R(f,\hat{\epsilon})$ and $\bar{C}_R(f,\hat{\epsilon})$ depend on the model assumed. The bounds we prove hold in a strong sense: all the lower bounds apply even if we combine the two random experiments into one — known to both communicators. All the upper bounds apply even if the experiments are separated (the reader can verify this to be the case in the examples given earlier in the section). Therefore, all bounds apply to both models.

We now prove lower bounds for randomized protocols and begin with error-free ones. In subsection 2.3, we showed that the 'largest f-constant rectangle' argument extends to average deterministic complexity. Here, we show that the bound $|S|/\lambda_f^S$, described in Theorem 2, holds also for the worst-case error-free (*Las Vegas*) complexity $\hat{C}_R(f,0)$.

Given any error-free deterministic protocol ϕ and any input (x,y), there exists an error-free deterministic protocol ϕ' that exchanges only two bits when the input is (x,y) and at most two bits more than ϕ for every other input. It is therefore not a priori clear that there must be an input with large expected communication ($\bar{L}_\Phi(x,y)$). We prove the lower bound by demonstrating a set S of inputs over which every deterministic protocol must exchange a large number of bits on the average.

We define the average complexity of a protocol Φ over a set $S \subseteq \{0,..,n-1\}\times\{0,..,n-1\}$ to be:

$$\bar{L}_\Phi^S \stackrel{\text{def}}{=} \frac{1}{|S|} \sum_{(x,y)\in S} \bar{L}_\Phi(x,y).$$

For deterministic ϕ's, $\mathcal{R}_\phi$ induces a partition $S_1,\ldots,S_J$ of S such that each S_j is the intersection of S with some rectangle $R_j \in \mathcal{R}_\phi$. All inputs in each S_j have the same codeword — $b_\phi(R_j)$ — whose length we denote by l_j. Therefore, $\bar{L}_\phi^S = \sum_{j=1}^{J} \frac{|S_j|}{|S|} l_j$. The l_j's are still lengths of codewords in a prefix-free set. An argument similar to that used in Theorem 8, can show that

Lemma 3 If S is partitioned by $\mathcal{R}_\phi$ into $S_1,\ldots,S_J$, then

$$\bar{L}_\phi^S \geq H\langle |S_j|/|S| : j=1,\ldots,J\rangle \quad \Box$$

Therefore, every protocol $\phi \in \Phi$ satisfies $\bar{L}_\phi^S \geq \log(|S|/\lambda_f^S)$ and from (3), $\bar{L}_\Phi^S \geq \log(|S|/\lambda_f^S)$. This implies the existence of at least one input with that complexity. Thus

Theorem 11 [OE 86] $\hat{C}_R(f,0) \geq \log(|S|/\lambda_f^S)$ for all $S \subseteq \{0,..,n-1\}\times\{0,..,n-1\}$ $\Box$

Example 7 Let $d_H^N(x,y)$ denote the Hamming distance between the two N-bit sequences x and y. For $N \geq 0$ and $0 \leq k \leq N$, define the function $H_k^N : \{0,1\}^N \times \{0,1\}^N \to \{0,1\}$ by

$$H_k^N(x,y) \stackrel{\text{def}}{=} \begin{cases} 1 & \text{if } d_H^N(x,y)=k \\ 0 & \text{otherwise.} \end{cases}$$

To determine the worst-case Las Vegas complexity of H_k^N, consider $\mathbf{1}_{H_k^N}$ (the set of inputs for which H_k^N is 1). Its size is $2^N\binom{N}{k}$. Yet, [PE 84] showed that for $0 \leq k \leq \lceil N/2 - \sqrt{N/4}\rceil$, the largest rectangle contained in $\mathbf{1}_{H_k^N}$ is of size $\binom{N}{k}$. Therefore, by a small modification of Lemma 3 and the argument used in Theorem 11, $\hat{C}_R(H_k^N, 0) \geq \lceil \log(2^N + 1) \rceil = N + 1$. This clearly implies that for all k's in the above range, $\hat{C}_R(H_k^N, 0) = N + 1$.

When n is a power of 2, the equality function is clearly equivalent to H_0^N. Therefore, $\hat{C}_R(equ, 0) = N + 1 = \log n + 1$ □

Theorem 11 can also be used to prove the randomized counterpart of Theorem 7:

Corollary 6 [OE 86] For all $n \leq s \leq n^2/2$, most functions in $\mathcal{F}_s$ have

$$\hat{C}_R(f, 0) \geq \log n - 4 \quad \square$$

In the rest of the section we prove lower bounds on randomized complexity when ϵ error is allowed. For each rectangle $R \in \{0,..,n-1\} \times \{0,..,n-1\}$, define its *impurity* to be:

$$\rho_f(R) \stackrel{\text{def}}{=} \min\left\{\frac{|R \cap \mathbf{1}_f|}{|R|}, \frac{|R \cap \mathbf{0}_f|}{|R|}\right\},$$

and for each $0 \leq \delta < \frac{1}{2}$, let

$$\lambda_f(\delta) \stackrel{\text{def}}{=} \max\{|R| : R \in \mathcal{R}_\phi \text{ and } \rho_f(R) < \frac{1}{2} - \delta\}$$

denote the size of the largest rectangle with impurity $< \frac{1}{2} - \delta$. In Section 2.3, we used λ_f, the size of the largest f-constant rectangle, to lower bound $\bar{C}_D(f, 0)$. Here, errors are allowed; therefore, f-constant rectangles have no special significance. We use $\lambda_f(\delta)$ instead.

Lemma 4 [OE 85] For every function $f \in \{0,1\}^{\{0,..,n-1\} \times \{0,..,n-1\}}$, protocol Φ (possibly randomized), and $\delta > 0$,

$$\bar{L}_\Phi \geq \left(1 - \frac{\bar{E}_\Phi(f)}{\frac{1}{2} - \delta}\right) \log \frac{n^2}{\lambda_f(\delta)}.$$

Proof: We first prove the lemma for deterministic protocols. Each deterministic protocol ϕ induces a partition $\mathcal{R}_\phi$ of $\{0,..,n-1\} \times \{0,..,n-1\}$ into rectangles (not necessarily f-constant since we allow errors). Let R be a rectangle in $\mathcal{R}_\phi$. From Corollary 3, ϕ assigns a unique value $v_\phi(R)$ to all inputs in R and therefore the average error $\bar{E}_\phi^R(f)$ that ϕ incurs over R satisfies:

$$\bar{E}_\phi^R(f) \stackrel{\text{def}}{=} \frac{1}{|R|} \sum_{(x,y) \in R} \chi_{\{(x,y): v_\phi(x,y) \neq f(x,y)\}}(x, y) \geq \rho_f(R).$$

Summing over all rectangles in $\mathcal{R}_\phi$:

$$\begin{aligned}\bar{E}_\phi(f) &\stackrel{\text{def}}{=} \frac{1}{n^2} \sum_{(x,y)\in\{0,..,n-1\}\times\{0,..,n-1\}} \chi_{\{(x,y):v_\phi(x,y)\neq f(x,y)\}}(x,y) \\ &= \frac{1}{n^2} \sum_{R\in\mathcal{R}_\phi} \sum_{(x,y)\in R} \chi_{\{(x,y):v_\phi(x,y)\neq f(x,y)\}}(x,y) \\ &\geq \sum_{R\in\mathcal{R}_\phi} \frac{|R|}{n^2}\rho_f(R) .\end{aligned}$$

Distinguishing rectangles with impurity $< \frac{1}{2} - \delta$ from those with impurity $\geq \frac{1}{2} - \delta$:

$$\begin{aligned}\bar{E}_\phi(f) &\geq \sum_{\{R\in\mathcal{R}_\phi:\rho_f(R)\geq\frac{1}{2}-\delta\}} \frac{|R|}{n^2}\rho_f(R) + \sum_{\{R\in\mathcal{R}_\phi:\rho_f(R)<\frac{1}{2}-\delta\}} \frac{|R|}{n^2}\rho_f(R) \\ &\geq \sum_{\{R\in\mathcal{R}_\phi:\rho_f(R)\geq\frac{1}{2}-\delta\}} \frac{|R|}{n^2}(\tfrac{1}{2}-\delta) .\end{aligned}$$

This inequality upper bounds the proportion of area occupied by rectangles with high impurity and can therefore be used to lower bound the proportion of area occupied by rectangles with low impurity:

$$\sum_{\{R\in\mathcal{R}_\phi:\rho_f(R)<\frac{1}{2}-\delta\}} \frac{|R|}{n^2} \geq 1 - \frac{\bar{E}_\phi(f)}{\frac{1}{2}-\delta} .$$

From the definition of $\lambda_f(\delta)$, $\{R \in \mathcal{R}_\phi : \rho_f(R) < \frac{1}{2} - \delta\} \subseteq \{R \in \mathcal{R}_\phi : |R| \leq \lambda_f(\delta)\}$. Therefore,

$$\sum_{\{R\in\mathcal{R}_\phi:|R|\leq\lambda_f(\delta)\}} \frac{|R|}{n^2} \geq 1 - \frac{\bar{E}_\phi(f)}{\frac{1}{2}-\delta} . \tag{5}$$

The sizes of rectangles in $\mathcal{R}_\phi$ and the average length of ϕ were related in Theorem 8 by:

$$\bar{L}_\phi \geq H\left\langle |R|/n^2 : R \in \mathcal{R}_\phi \right\rangle . \tag{6}$$

A simple information-theoretic argument can be used to show that for every probability distribution $\langle p_j \;:\; j = 1, \ldots, J\rangle$, and every $0 \leq A,\ r \leq 1$,

$$\sum_{\{j:p_j\leq r\}} p_j \geq A \quad\Rightarrow\quad H\langle p_j \;:\; j = 1, \ldots, J\rangle \geq A \log\frac{1}{r}. \tag{7}$$

Combining (5), (6), and (7):

$$\bar{L}_\phi \geq \left(1 - \frac{\bar{E}_\phi(f)}{\frac{1}{2}-\delta}\right) \log\frac{n^2}{\lambda_f(\delta)}.$$

This inequality can be easily extended to randomized protocols. From (4), every randomized protocol Φ satisfies: $\bar{L}_\Phi = \sum_\phi p(\phi)\bar{L}_\phi$ and $\bar{E}_\Phi(f) = \sum_\phi p(\phi)\bar{E}_\phi(f)$. Thus,

$$\bar{L}_\Phi \geq \left(1 - \frac{\bar{E}_\Phi(f)}{\frac{1}{2}-\delta}\right) \log\frac{n^2}{\lambda_f(\delta)} \quad \square$$

The inequality proved in Lemma 4 clearly holds if we replace $\bar{E}_\Phi(f)$ by the worst-case error $\hat{E}_\Phi(f)$. Therefore, we can lower bound $\bar{C}_R(f,\hat{\epsilon})$ by showing that $\lambda_f(\delta)$ is small. In the next lemma and Theorem 12 we do so for most functions. In Lemma 6 and Corollary 7, we prove the same results for the inner-product function.

Lemma 5 [OE 85] For every $0 < \delta < \frac{1}{2}$, most functions in $\{0,1\}^{\{0,..,n-1\}\times\{0,..,n-1\}}$ have $\lambda_f(\delta) < \frac{n}{\delta^2}$.

Proof: Fix a rectangle R of area $|R|$ in $\{0,..,n-1\}\times\{0,..,n-1\}$. The number of functions in $\{0,1\}^{\{0,..,n-1\}\times\{0,..,n-1\}}$ with impurity $\rho_f(R) < \frac{1}{2} - \delta$ is

$$\begin{aligned}
2\cdot\sum_{k=0}^{|R|(\frac{1}{2}-\delta)}\binom{|R|}{k}\cdot 2^{n^2-|R|} &\le 2|R|(\tfrac{1}{2}-\delta)\binom{|R|}{|R|(\frac{1}{2}-\delta)}\cdot 2^{n^2-|R|}\\
&\le 2|R|(\tfrac{1}{2}-\delta)\sqrt{\frac{1}{\pi(\frac{1}{2}-\delta)(\frac{1}{2}+\delta)|R|}}\cdot 2^{|R|h(\frac{1}{2}-\delta)}2^{n^2-|R|}\\
&< \sqrt{\frac{2(\frac{1}{2}-\delta)|R|}{\pi(\frac{1}{2}+\delta)}}\cdot 2^{|R|(1-\frac{2\delta^2}{\ln 2})}\cdot 2^{n^2-|R|}\\
&\le \sqrt{|R|}\cdot 2^{-|R|\frac{2\delta^2}{\ln 2}}\cdot 2^{n^2}.
\end{aligned}$$

Therefore, for any $A \le n^2$, the fraction of functions that have $\rho_f(R) < \frac{1}{2} - \delta$ for some rectangle R of area $|R| \ge A$, is at most $\sqrt{n}\cdot 2^{-\frac{2\delta^2 A}{\ln 2}}\cdot 2^{2n}$. Taking $A = n/\delta^2$, a fraction of at least $1 - n2^{-.88n}$ of the functions in $\{0,1\}^{\{0,..,n-1\}\times\{0,..,n-1\}}$ have $\lambda_f(\delta) \le n/\delta^2$ $\square$

Substituting $1/\log n$ for δ in Lemma 5 and employing Lemma 4, we obtain:

Theorem 12 [OE 85,CG 85] Most functions in $\{0,1\}^{\{0,..,n-1\}\times\{0,..,n-1\}}$ have

$$\bar{C}_R(f,\hat{\epsilon}) \ge (1-2\epsilon)(\log n - 2\log\log n) - 2 \quad \square$$

Note that the $1 - 2\epsilon$ term is necessary because with probability 2ϵ, $P_\mathcal{X}$ and $P_\mathcal{Y}$ can guess the result and be wrong with half that probability.

As for the inner-product function, [PE 84] proved that $\hat{C}_R(i.p.,\hat{\epsilon}) \ge C(\epsilon)\cdot\log n$ for $\epsilon < 1/8$. In [CG 85], an elegant argument of J.H. Lindsey ([ES 74], page 88) was extended to improve the bound. This argument uses the Hadamard structure of the function matrix and is described next.

Let A be an $n\times n$ matrix with elements in $\{-1,+1\}$. denote the xth row of A (regarded as a row vector) by $r_x(A)$ and denote its transpose by $r_x^t(A)$. Call A *Hadamard* if its rows are *orthogonal*, i.e., $r_{x'}(A)\cdot r_{x''}^t(A) = 0$ for all $x' \ne x''$. (When $x' = x''$, $r_{x'}(A)\cdot r_{x''}^t(A) = n$ because $A(x,y) \in \{-1,+1\}$.)

To show that the function matrix of the modified inner-product function, $M_{i.p.^\star}$, is Hadamard, note that if $x' \neq x''$ then there exists an index i such that the ith bits in the binary representations of x' and x'' differ. For every integer y, let $y^{c(i)}$ denote the integer whose binary representation is obtained from that of y by complementing the ith bit. Then $i.p.^\star(x',y)\cdot i.p.^\star(x'',y) + i.p.^\star(x',y^{c(i)})\cdot i.p.^\star(x'',y^{c(i)}) = 0$. Summing this equality over all integers y whose ith bit is 0, we obtain: $r_{x'}(M_{i.p.^\star})\cdot r^t_{x''}(M_{i.p.^\star}) = 0$.

Lemma 6 [ES 74,CG 85] For any rectangle $|R| \in \{0,..,n-1\}\times\{0,..,n-1\}$,

$$|\sum_{(x,y)\in R} i.p.^\star(x,y)| \leq \sqrt{n|R|}.$$

Proof: We show that for any $n\times n$ Hadamard matrix H and any rectangle $R = R_{\mathcal{X}}\times R_{\mathcal{Y}} \subseteq \{0,..,n-1\}\times\{0,..,n-1\}$, $\quad |\sum_{\{(x,y)\in R\}} H(x,y)| \leq \sqrt{n|R|}$.

Let $1_{R_{\mathcal{Y}}}$ be the n-dimensional row vector whose yth coordinate is 1 if $y \in R_{\mathcal{Y}}$ and 0 otherwise. Let $1^t_{R_{\mathcal{Y}}}$ denote its transpose. Using the Cauchy Schwarz Inequality and orthogonality:

$$\begin{aligned}
|\textstyle\sum_{(x,y)\in R} H(x,y)| &= |\textstyle\sum_{x\in R_{\mathcal{X}}}\sum_{y\in R_{\mathcal{Y}}} H(x,y)| \\
&= |\textstyle\sum_{x\in R_{\mathcal{X}}} r_x(H)\cdot 1^t_{R_{\mathcal{Y}}}| \\
&\leq ||\textstyle\sum_{x\in R_{\mathcal{X}}} r_x(H)||\cdot||1_{R_{\mathcal{Y}}}|| \\
&= \sqrt{\textstyle\sum_{x\in R_{\mathcal{X}}} ||r_x(H)||^2}\cdot||1^t_{R_{\mathcal{Y}}}|| \\
&= \sqrt{n|R|} \quad \Box
\end{aligned}$$

The Lemma clearly implies that $\rho_{i.p.^\star}(R) \geq \frac{1}{2} - \sqrt{\frac{n}{|R|}}$. Therefore, $\lambda_{i.p.^\star}(\delta) \leq \frac{n}{\delta^2}$ and, as in Theorem 12:

Corollary 7 [PE 84,CG 85] $\quad \bar{C}_R(i.p., \hat{\epsilon}) \geq (1-2\epsilon)(\log n - \log\log n) - 2 \quad \Box$

We note that with a more careful analysis, the bound can be further improved. Few other bounds are known for specific functions. The *set intersection function*:

$$s.i.(x,y) \stackrel{\text{def}}{=} \begin{cases} 1 & \text{if for some } i,\ x_i = y_i \\ 0 & \text{if for all } i,\ x_i \neq y_i \end{cases}$$

(where, again, x and y are regarded as a $\log n$ bit sequences, $x_1,\ldots,x_{\log n}$ and $y_1,\ldots,y_{\log n}$) was shown by [BFS 86] to have $\hat{C}_R(s.i., \epsilon) \geq \Omega(\sqrt{\log n})$ and [Yao 83] showed that the non boolean function $(x\cdot y) \bmod p$ has complexity $\Theta(k)$ when p is a k bit prime.

What about sparse functions? For average complexity, we saw in Theorem 9 that for all functions, $\bar{C}_D(f,0) \le \log \frac{1_f}{n} + 8.3 \log\log n + c$. Since, $\bar{C}_R(f,\hat{\epsilon}) \le \bar{C}_D(f,0)$ for all functions f, the same inequality holds for $\bar{C}_R(f,\hat{\epsilon})$. But can we further reduce the complexity by allowing errors and randomization? The next theorem (part 1) shows that for most functions we cannot. As for worst-case complexity, Corollary 6 showed that for all $n \le s \le n^2/2$, most functions in $\mathcal{F}_s$ have $\hat{C}_R(f,0) \ge \log n - 2$. Can allowing errors reduce the complexity now? Part 2 of the next theorem says that they can.

Theorem 13 [OE 85] For all $0 < \epsilon \le \frac{1}{2}$, and $s \ge n \ge 2$, *most* functions in $\mathcal{F}_s$ have:

1) $\bar{C}_R(f,\hat{\epsilon}) \ge (1-2\epsilon)\left(\log \frac{s}{n} - 2\log\log \frac{s}{n}\right) - 9$

2) $\hat{C}_R(f,\hat{\epsilon}) \le 2 + (1-2\epsilon)\left(\log \frac{s}{n} + 5.3 \log\log n + 2\log \frac{1}{\epsilon} + c\right)$ □

4 Complexities of Most Functions

In the two preceding sections, we defined six complexity measures: $\hat{C}_D(f,0)$, $\bar{C}_D(f,0)$, $\hat{C}_R(f,0)$, $\bar{C}_R(f,0)$, $\hat{C}_R(f,\hat{\epsilon})$, and $\bar{C}_R(f,\hat{\epsilon})$. It can be shown that for all functions f, $\bar{C}_R(f,0) = \bar{C}_D(f,0)$ leaving five different measures that we classify in this section.

Remark: As for randomized complexities, one could define the deterministic measures $\bar{C}_D(f,\hat{\epsilon})$ and $\hat{C}_D(f,\hat{\epsilon})$. However, for deterministic protocols, $\hat{E}_\phi(f)$ is either 0 or 1, thus $\hat{C}_D(f,\hat{\epsilon}) = \hat{C}_D(f,0)$ and $\bar{C}_D(f,\hat{\epsilon}) = \bar{C}_D(f,0)$.

We say that complexity measure A is *smaller* than a complexity measure B if $A(f) \le B(f)$ for all functions f. For example, $\bar{C}_R(f,\hat{\epsilon}) \le \hat{C}_R(f,\hat{\epsilon})$ for all functions f. This relation, forms a partial order on the five complexity measures which is depicted in Figure 1.

As in section 2.2, let $\mathbf{1}_f$ be the set of inputs for which f is 1 and let $\mathcal{F}_s \stackrel{\text{def}}{=} \{f : |\mathbf{1}_f| = s\}$ be the set of functions with s ones in their function matrix. There are $\binom{n^2}{s}$ functions in $\mathcal{F}_s$ and without loss of generality, we consider only functions with less ones than zeros, that is, $\mathbf{1}_f \le n^2/2$. We stated earlier that for every fixed integer s between n and $n^2/2$:

1) [Corollary 6] most functions in $\mathcal{F}_s$ have worst case error-free (Las Vegas) complexities, $\hat{C}_D(f,0)$ and $\hat{C}_R(f,0)$ at least $\ge \log n - 4$ (even when $s = n$).

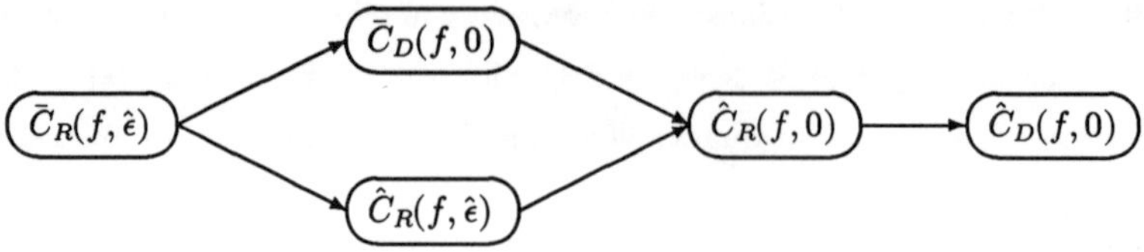

Figure 1: Partial order on the complexity measures

2) [Theorem 9] All functions in $\mathcal{F}_s$ have $\bar{C}_D(f,0) \leq \log \frac{s}{n} + 8.3 \log\log n + c$ (that is, a deterministic protocol that incurs no errors and, on the average, exchanges $\leq \log \frac{s}{n} + 8.3 \log\log n + c$ bits).

3) [Theorem 13, part 2] Most of the functions in $\mathcal{F}_s$ have $\hat{C}_R(f,\hat{\epsilon}) \leq 2 + (1-2\epsilon)(\log \frac{s}{n} + 5.3 \log\log n + 2\log \frac{1}{\epsilon} + c)$.

4) [Theorem 13, part 1] For most of the functions in $\mathcal{F}_s$, the bounds in (2) and (3) are tight: $\bar{C}_R(f,\hat{\epsilon}) \geq (1-2\epsilon)(\log \frac{s}{n} - 2\log\log \frac{s}{n}) - 9$.

Superimposing these bounds on the partial order of figure 1, we obtain Figure 2. Lower bounds are drawn as dashed vertical lines and upper bounds as dotted vertical lines. For example, the leftmost line is a lower bound derived from the bound in (4) and the lower dotted line at the center of the figure is derived from the bound in (3). Since complexities in the figure increase from left to right, these two lines (bounds) imply that for most functions in $\mathcal{F}_s$, both $\hat{C}_R(f,\hat{\epsilon})$ and $\bar{C}_R(f,\hat{\epsilon})$ are about $(1-2\epsilon)\log \frac{s}{n}$.

It is apparent from the figure that:

1) All complexity measures fall in two very distinct classes. One (consisting of $\hat{C}_D(f,0)$ and $\hat{C}_R(f,0)$) is at least $\log n - 4$ bits for every s. The other (consisting of $\bar{C}_D(f,0)$, $\hat{C}_R(f,\hat{\epsilon})$, and $\bar{C}_R(f,\hat{\epsilon})$) decreases with s and is about $\log \frac{s}{n}$ bits. The difference between the two classes ranges from a constant factor (for $s \approx \frac{n^2}{2}$) to exponential (for $s \approx n$ or $s \approx n \log n$).

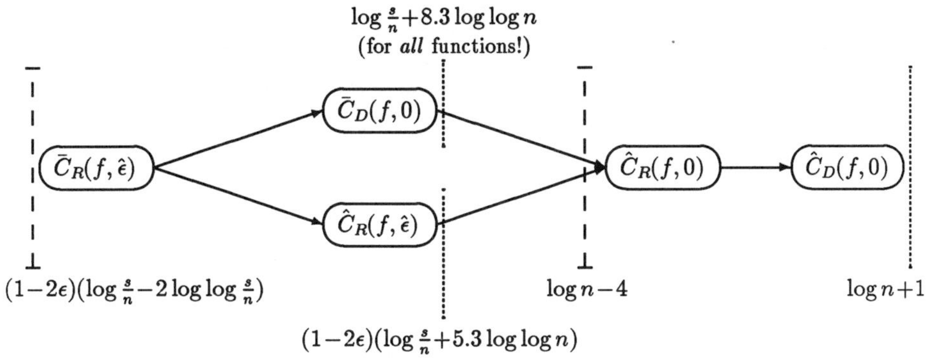

Figure 2: Bounds for most functions in $\mathcal{F}_s$ $(n \le s \le n^2/2)$

For most functions, $|\mathbf{1}_f| \approx \frac{n^2}{2}$. Therefore, all complexity measures coincide and are about $\log n$ bits. However, for most sparse functions ($|\mathbf{1}_f| \ll \frac{n^2}{2}$):

2) In error-free computation, considerably fewer bits are needed on the average than at worst (most interesting, of course, is the difference between $\hat{C}_D(f,0)$ and $\bar{C}_D(f,0)$). This is one of the few cases in complexity theory where such a discrepancy can be proved.

3) In error-free computation, randomization does not help reduce the worst-case complexity. Even for $s \approx n$, most functions in $\mathcal{F}_s$ have $\hat{C}_R(f,0) \ge \log n - 4$. Hence, Las Vegas is not much better than determinism, and the function described in [MS 82] (see Example 6) is an exception.

4) However, if errors are allowed, randomization does reduce the worst-case complexity. In fact, in this case, it is close to average error-free deterministic complexity. Thus the behavior of the equality function is typical of functions in its class: $\mathcal{F}_s$.

5 Related Results

In this section, we review results related to communication complexity. We describe the first two in detail and only briefly discuss the rest.

5.1 Nondeterministic Communication Complexity

Nondeterministic communication protocols are the communication complexity counterparts of Nondeterministic Turing Machines (see [HU 79]). Each communicator can base his transmissions on guesses that he makes along the way. The protocol is said to *compute* f *nondeterministically* (or, to be a *nondeterministic protocol for* f) if for all inputs x, y with $f(x, y) = 0$ the computation always yields 0; and whenever $f(x, y) = 1$ there exists at least one possible guess-sequence that results in a computed 1. Note that a nondeterministic protocol for f is not necessarily a nondeterministic protocol for the complement $\overline{f}$ of f.

We define the worst-case complexity of a nondeterministic protocol as the largest number of transmitted bits, maximized over all inputs and all possible guesses. The nondeterministic complexity $\hat{C}_{ND}(f)$ of a function f is defined as the smallest worst-case complexity over all nondeterministic protocols for f.

Example 8 Let n be a power of 2 and consider the complement of the equality function defined for $x, y \in \{0,..,n-1\}$ by:

$$\overline{equ}(x, y) \stackrel{\text{def}}{=} \begin{cases} 1 & \text{if } x \neq y \\ 0 & \text{if } x = y \end{cases} .$$

It was shown in section 2.2 that $\hat{C}_D(\overline{equ}, 0) = \hat{C}_D(equ, 0) = \lceil \log n \rceil + 1$. The following protocol shows that $\hat{C}_{ND}(\overline{equ}) \leq 2 + \lceil \log \log n \rceil$ (again, each integer $x \in \{0,..,n-1\}$ is regarded as a $\lceil \log n \rceil$ bits long sequence: $x_1, \ldots, x_{\log n}$).

1) $P_{\mathcal{X}}$ guesses an integer $1 \leq i \leq \lceil \log n \rceil$ and transmits i and X_i.

2) $P_{\mathcal{Y}}$ responds by transmitting 0 if $X_i = Y_i$ and 1 otherwise.

3) They take the bit transmitted by $P_{\mathcal{Y}}$ to be the computed value.

Clearly, if $\overline{equ}(X, Y) = 0$ (that is, $X = Y$), $P_{\mathcal{Y}}$ will always transmit 0. If $\overline{equ}(X, Y) = 1$, there must be at least one i such that $X_i \neq Y_i$. If this i is picked, the decided value of the function will be 1. Therefore, the protocol described above computes $\overline{equ}$ and the number of bits transmitted is always $2 + \log\log n$. Once a few observations are made, it is easy to see that this is the best possible bound and that nondeterminism cannot help reduce the complexity of the equality function $\square$

A set $\mathcal{R}$ of, possibly overlapping, rectangles is called a *cover* for a set $S \subseteq \{0,..,n-1\} \times \{0,..,n-1\}$ (or called a cover for S) if $\cup\{R \in \mathcal{R}\} = S$. [LS 81] showed that in the same way that a deterministic protocol for f corresponds to a partition (i.e., to a nonoverlapping cover) of $\{0,..,n-1\} \times \{0,..,n-1\} = \mathbf{1}_f \cup \mathbf{0}_f$ into f-constant rectangles, a nondeterministic protocol for f corresponds to a (possibly overlapping) cover of the set $\mathbf{1}_f \stackrel{\text{def}}{=} \{(x,y) : f(x,y) = 1\}$.

The crux of the lower-bound techniques of Section 2.2 was the inequality $\hat{C}_D(f,0) \geq \log \mu_d(f)$. In fact, there is an even closer relation between the smallest number of (possibly overlapping) rectangles needed to cover $\mathbf{1}_f$, which we denote by $\mu_o(f)$, and $\hat{C}_{ND}(f)$:

Theorem 14 [LS 81,JKS 84] $\qquad \log \mu_o(f) \leq \hat{C}_{ND}(f) \leq \log(\mu_o(f) + 1) \qquad \square$

Moreover, as noted in [JKS 84], the upper bound holds even for the restricted one-way-communication model where $P_{\mathcal{X}}$ transmits a message and $P_{\mathcal{Y}}$ responds with the value of the function.

Since n rectangles are needed to cover $\{(x,y) : x = y\}$, the theorem clearly implies the claim made in the last example. For some functions, however, both f and $\bar{f}$ are easier nondeterministically than deterministically.

Example 9 [MS 82] The function *c.e.* of Example 6 was shown to have $\hat{C}_D(c.e.,0) = M^2$. The following protocols show that $\hat{C}_{ND}(c.e.,0) \leq M + \log M$ and $\hat{C}_{ND}(\overline{c.e.},0) \leq M \log M$.

Protocol for *c.e.*:

1) $P_{\mathcal{X}}$ picks an integer $1 \leq i \leq M$ and transmits i and X^i.
2) $P_{\mathcal{Y}}$ transmits 1 if $X^i = Y^i$ and 0 otherwise.
3) They take the bit transmitted by $P_{\mathcal{Y}}$ to be the computed value.

Protocol for $\overline{c.e.}$:

1) For $i = 1, \ldots, M$,

i) $P_\mathcal{X}$ guesses an integer $j(i)$ and transmits $X^i_{j(i)}$.

ii) $P_\mathcal{Y}$ transmits $Y^i_{j(i)}$.

2) The computed value is 1 if $X^i_{j(i)} = Y^i_{j(i)}$ for some i and is 0 otherwise □

One of the most interesting results concerning nondeterministic communication complexity states that this is almost the maximal gap possible: there is never more than a quadratic difference between deterministic and nondeterministic communication complexity. Recall the discussion following Theorem 5. We noted that the result holds even for overlapping rectangles. Namely, $\hat{C}_D(f,0) \le O(\log^2(\mu_o(f) + \mu_o(\bar{f})))$. Combined with the last theorem, this implies that

Theorem 15 [AUY 83] $\hat{C}_D(f,0) \le O\left[\max^2(\hat{C}_{ND}(f), \hat{C}_{ND}(\bar{f}))\right]$ □

This is a much sought-after type of result in computational complexity. The corresponding result would imply that if f is in both NP and co-NP, then it is in P.

The 'converse' of this theorem is also very intriguing. We mentioned the interesting open question of whether the bound of Theorem 5 is tight. M. Yanakakis noted that, as far as overlapping covers are concerned, it almost is. For the component equality function, $\log(\mu_o(c.e.) + \mu_o(\overline{c.e.})) \le M \log M$, yet $\hat{C}_D(c.e.,0) = M^2$.

For additional analogies between complexity classes in computation and communication, including BPP, PSPACE, PP, Σ_k, and Π_k, see [BFS 86].

5.2 Exchanging Related Information

So far, we considered computing functions with the smallest possible range: 2 (or 1) values. On the other extreme are functions with the largest possible range: n^2 values. Clearly, all these functions are equivalent to computing $f(x,y) \stackrel{\text{def}}{=} (x,y)$ and therefore amount to *exchanging* (or *sharing*) X and Y. However, in the setup described earlier where all inputs are effectively equally likely, $P_\mathcal{X}$ has to describe X and $P_\mathcal{Y}$ has to describe Y, hence $2\lceil \log n \rceil$ bits have to be transmitted both on the average and at worst. To add meaning to the problem, assume that X and Y are random variables distributed according to some joint probability distribution $p(x,y)$ over a finite product set $S_\mathcal{X} \times S_\mathcal{Y}$. Let $\bar{C}(p)$ (respectively, $\hat{C}(p)$) be the number of bits that $P_\mathcal{X}$ and $P_\mathcal{Y}$ must transmit on the average (resp., at worst) in order to exchange X and Y.

The *entropy of a random variable* is defined as the entropy of its underlying probability distribution. Therefore,

$$H(X,Y) \stackrel{\text{def}}{=} H\langle p(x,y) : (x,y) \in S_{\mathcal{X}} \times S_{\mathcal{Y}} \rangle = \sum_{(x,y) \in S_{\mathcal{X}} \times S_{\mathcal{Y}}} p(x,y) \log \frac{1}{p(x,y)}.$$

Let $p(y) \stackrel{\text{def}}{=} \sum_{x \in S_{\mathcal{X}}} p(x,y)$ denote the *marginal probability distribution of* Y and define the *conditional entropy of* X *given* Y as:

$$H(X|Y) \stackrel{\text{def}}{=} \sum_{y \in S_{\mathcal{Y}}} p(y) H(X|Y=y).$$

It was shown in [EO 84] that for all probability distributions,

$$H(X|Y) + H(Y|X) \leq \bar{C}(p) \leq H(X,Y) + 2$$

with the upper bound achievable using two Huffman Codes. For an arbitrary probability distribution the bounds are not tight. However, as demonstrated by the next example, given only the entropies, these are the best possible bounds: there are probability distributions where $\bar{C}(p)$ is close to the lower bound and distributions where $\bar{C}(p)$ is almost the upper bound.

Example 10 The probability distribution defined by

$$p_n(x,y) \stackrel{\text{def}}{=} \begin{cases} 1/n & \text{if } 0 \leq x = y < n \\ 0 & \text{otherwise} \end{cases}$$

has $H(X,Y) = \log n$ and $H(X|Y) + H(Y|X) = 0$ displaying unbounded discrepancy between the lower and upper bounds. Clearly, $\bar{C}(p_n) = \hat{C}(p_n) = 0$ which is the lower bound.

By continuity, for small δ, the probability distribution

$$q_{\delta,n}(x,y) \stackrel{\text{def}}{=} \begin{cases} (1-\delta)/n & \text{if } 0 \leq x = y < n \\ \delta/(n^2 - n) & \text{otherwise} \end{cases}$$

has $H(X,Y) \approx \log n$ and $H(X|Y) + H(Y|X) \approx 0$. Yet, it follows from the next paragraph that, $\bar{C}(q_{\delta,n}) \approx \log n$, i.e., close to the upper bound $\square$

Therefore, to derive more meaningful bounds, we need additional information about the probability distribution. A reasonable assumption is that $p(x,y) > 0$ for all $(x,y) \in$

$S_\mathcal{X} \times S_\mathcal{Y}$. One would hope that if X and Y satisfy some regularity constraints, then the lower bound can be achieved ([SW 73]). At least, that is the case in traditional information theory. Unfortunately, it can be shown that since no errors are allowed, all probability distributions satisfying $p(x,y) > 0$ for all x and y have $\bar{C}(p)$ at most 2 bits less than the upper bound:

$$H(X,Y) \le \bar{C}(p) \le H(X,Y) + 2.$$

Therefore any reduction in communication complexity must come from inputs with $p(x,y) = 0$. That is, p is 0 outside some *support set*, S_p, which is strictly contained in $S_\mathcal{X} \times S_\mathcal{Y}$. The case where $p(x,y)$ is uniform over S_p was explored in [EO 84] and it was shown that for all such probability distributions, the lower bound can almost be achieved:

$$\bar{C}(p) \le H(X|Y) + H(Y|X) + 3.1 \log\log \max(|S_\mathcal{X}|, |S_\mathcal{Y}|) + c.$$

It should be noted that for most such distributions, $H(X|Y) + H(Y|X)$ is in the order of $\log |S_\mathcal{X}|$.

Example 11 [Due to T. Cover] $P_\mathcal{X}$ has a sequence X of N bits and $P_\mathcal{Y}$ has a sequence Y of N bits. The two sequences are known to be a cyclic shift of each other. How many bits must be transmitted in the worst case for each person to find the other's sequence?

Note that we did not specify the underlying probability distribution p. Instead, we described its *support set*: $S_p \stackrel{\text{def}}{=} \{(x,y) : p(x,y) > 0\}$ It can be shown that $\hat{L}(p)$ depends on p only via S_p. Therefore, we can take the distribution that yields the best bound amongst probability distributions with $S_p = \{(x,y) : x \text{ is a cyclic shift of } y\}$. Taking p_0 to be uniformly distributed over this set, we get $\hat{C}(p) = \hat{C}(p_0) \ge \bar{C}(p_0) \ge H(X|Y) + H(Y|X) \ge 2\log N - 1$. The number of bits transmitted according to the following protocol is within 3 bits of the lower bound: Let Z be the lexicographically largest sequence among all cyclic shifts of X. Then Z is also the largest sequence among all cyclic shifts of Y. Both $P_\mathcal{X}$ and $P_\mathcal{Y}$ can find Z. Therefore, $P_\mathcal{X}$ transmits to $P_\mathcal{Y}$ the number of times Z should be cyclicly right shifted to obtain X ($\lceil \log N \rceil$ bits) and $P_\mathcal{Y}$ does the same $\square$

Example 12 [EO 84] $P_\mathcal{X}$ has a sequence X of N bits and $P_\mathcal{Y}$ has a sequence Y of N bits. The two sequences are known to be different in at most K locations ($d_H(X,Y) \le K$).

How many bits must be transmitted in the worst case for each person to find the other's sequence?

Here, $S_p = \{(x,y) \in \{0,1\}^N : d_H(x,y) \leq K\}$. Taking p_0 to be the uniform distribution over this set, we obtain $\hat{C}(p_0) \geq 2\log\left(\sum_{k=0}^{K}\binom{N}{k}\right)$ for all distributions with the above support. A general protocol described in [EO 84] shows that $\hat{C}(p_0) \leq 2\log\left(\sum_{k=0}^{K}\binom{N}{k}\right) + \log N$ which holds for all distributions with the same support. The two bounds are asymptotically tight for every fixed K. Moreover, for Ks growing with N, the ratio between the upper and lower bounds approaches 1. However, for fixed values of K, there is a (very small) ratio between the two. For $K = 1,2,3$ the ratios are 1.5, 1.25, 1.166 respectively. The following protocols achieve worst-case complexities of $2(1+\log N)$ for $K = 1$, and $2K \cdot \log N$ for $K = 2,3$ thus reducing the ratio between the upper and lower bounds to 1 for these cases. (For simplicity, we assume that N is a power of 2).

$K = 1$. If $N = 1$ then exchanging X, Y achieves $2 \cdot (1+\log N)$. Assume that for sequences of length $N/2$ the protocol transmits at most $2 \cdot (1+\log N/2)$ bits. Given a sequence of length N, $P_\mathcal{X}$ transmits to $P_\mathcal{Y}$ the parity of the first $N/2$ bits of X and $P_\mathcal{Y}$ transmits to $P_\mathcal{X}$ the parity of the first $N/2$ bits of Y. If the parities differ, $P_\mathcal{X}$ and $P_\mathcal{Y}$ know that there exists $0 \leq i < N/2$ such that $X_i \neq Y_i$ and they use the $N/2$ protocol on the subsequences $\langle X_i\rangle_{i=0}^{N/2-1}$ and $\langle Y_i\rangle_{i=0}^{N/2-1}$. If the parities are the same, there is at most one i, $N/2 \leq i < N$ such that $X_i \neq Y_i$ so they use the protocol on the subsequences $\langle X_i\rangle_{i=N/2}^{N-1}$ and $\langle Y_i\rangle_{i=N/2}^{N-1}$. In either case, the total number of bits is at most $2 + 2\cdot(1+\log N/2) = 2\cdot(1+\log N)$.

$K = 2$. For $m = 1,\ldots,\log N$, let $A_m \stackrel{\text{def}}{=} \{i : 0 \leq i < N$ and the mth least significant bit in the binary representation of i is $1\}$. For $m = 1,\ldots,\log N$, $P_\mathcal{X}$ transmits $\bigoplus_{i\in A_m} X_i$ to $P_\mathcal{Y}$ and $P_\mathcal{Y}$ transmits $\bigoplus_{i\in A_m} Y_i$ to $P_\mathcal{X}$ ($\bigoplus$ denotes exclusive or). Let B be the $\log N$ bit long binary number whose mth least significant bit is 1 *iff* the parities corresponding to A_m are different. If all parities are the same ($B = 0$) then either $X = Y$ or $X_0 \neq Y_0$ so $P_\mathcal{X}$ and $P_\mathcal{Y}$ exchange the 0th bit to know which is the case. If the parities are not all equal, let M be any integer such that the parities for A_M differ (the Mth least significant bit of B is one). There is at most one $i \notin A_M$ such that $X_i \neq Y_i$ so $P_\mathcal{X}$ and $P_\mathcal{Y}$ use the scheme for $K = 1$ described above on the subsequences $\langle X_i\rangle_{i\notin A_M}$ and $\langle Y_i\rangle_{i\notin A_M}$. If they find that the two subsequences are equal then $X_B \neq Y_B$ and all other bits are the same. If they find that $X_C \neq Y_C$ say, then also, $X_{C\oplus B} \neq Y_{C\oplus B}$ and all the other bits are equal. The total number of bits exchanged is $4 \log N$.

$K = 3$. An easy combination of the above protocols results in worst-case length of at most $6 \cdot \log N$ bits □

5.3 Additional Results

We now briefly survey other results, related to communication complexity. They are described in order of relevance to the model discussed in the previous sections.

Unbounded-Error Communication Complexity

[PtS 84] considered randomized communication complexity where the only requirement is that the probability of error for each input be less than $\frac{1}{2}$ (but can be arbitrarily close). The complexity of a protocol is defined as the largest number of transmitted bits maximized over all inputs and all outcomes of the random experiments (taking the expectation over random experiments would trivialize the problem here). The *unbounded-error communication complexity* of a function f is then defined as the minimum complexity, over all protocols that compute f with probability of error less than $\frac{1}{2}$ for each input.

[PtS 84] showed that under these requirements, there is at most one bit difference between one and two-way communication complexity. Using this observation, they demonstrated a function with high unbounded-error complexity. Their results were extended by [AFR 85] who showed that the unbounded-error communication complexity of most functions is at least $\log n - 5$. Note that this result is, on the one hand, stronger than Theorem 12 as it holds for unbounded errors. On the other hand, it is weaker than Theorem 12 since the complexity is defined as the largest number of bits ever transmitted rather than as the average over the random experiments.

Arbitrary Partition of Inputs

Several papers, including [LS 81], [Yao 81], [PpS 82], [DGS 84], and [JK 84], viewed f as a boolean function defined on $\{0,1\}^{2N}$. They assumed that $P_{\mathcal{X}}$ knows N of the input bits and $P_{\mathcal{Y}}$ knows the rest. The communication complexity of f is defined as the smallest number of bits that the communicators need to transmit, minimized over all such partitions of the input bits. It was shown that most functions still require transmission of almost N bits, even with the most favorable partition of input bits. This formulation of the problem was motivated by lower-bound proofs on the area-time complexity of VLSI chips in the model introduced by [Th 80].

K-Round Protocols

[PpS 82] called protocols with at most K messages — K-round protocols. They demonstrated a function with a 3-round protocol of worst-case length $\leq 1 + 2\log N$ but for which every 2-round protocol requires at least $\sqrt{N}/2\log N$ bits. They conjectured that for every K, an extension of this function requires $O[(K-1)\log N]$ bits using K-round protocols but $\Omega(N^{1/c})$ bits using $(K-1)$-round protocols (where c is a constant). This was proved by [DGS 84] who showed that $\Omega(\sqrt{N}/K^4\log^3 N)$ bits are needed.

Communication Complexity in the Presence of Noise

[Ja' 85] considered communication complexity when the communication channel commits δ error in transmitting each bit. He showed (with A. Odlyzko) that for all $\delta < \frac{1}{2}$, there is a protocol that transmits at most $\log n + o(\log n)$ bits and computes the equality function with probability of error $\leq 1/cn$ for some constant c. This is contrasted with decoding (discovering the random integer known to the other communicator) where a basic information-theoretic result of [Sh 48] says that at least $\frac{1}{1-H(\delta)}\log n$ bits must be transmitted to ensure diminishing probability of error. [Ja' 85] also demonstrated a boolean function for which ensuring probability of error decreasing faster than $1/\log n$, requires at least $(1+g(\delta))\log n$ bits for some function g.

Communicating the Value of a Function

[AC 81] were interested in a problem where $P_{\mathcal{X}}$ having a sequence X of n (possibly dependent) random variables and $P_{\mathcal{Y}}$ having a sequence Y, communicate noninteractively, in a one-way fashion, to a third person who wants to compute $f(X,Y)$ with an arbitrarily small probability of error. They demonstrated a class of functions for which at least $n(H(X|Y)+H(Y|X))$ bits must be communicated. A simple proof of this result was given in [ElG 83].

Communication Complexity for Continuous Functions

In this manuscript, we dealt with discrete functions. Even prior to the introduction of this model, [Ab 78] defined a communication complexity measure for continuous functions. In his model, $\mathcal{R}$ is the set of reals, $f : \mathcal{R}^a \times \mathcal{R}^b \rightarrow \mathcal{R}$ has continuous first and second derivatives, $P_{\mathcal{X}}$ knows $X \in \mathcal{R}^a$, $P_{\mathcal{Y}}$ knows $Y \in \mathcal{R}^b$, and they can evaluate only functions with continuous first and second derivatives. How many *real numbers* do they need to exchange to find $f(x,y)$? Defining the a by b matrix $\Delta(i,j) \stackrel{\text{def}}{=} \partial^2 f/\partial x_i \partial y_j$, Abelson showed that at least rank(Δ) real numbers must be transmitted.

Communicating via a Chain of Processors

[Ti 84] looked at two processors communicating via a chain of $p-1$ intermediate processors (p links) in order to compute a function $f(x,y)$. Extending the rank and fooling-set methods described in Section 2.1, he showed that $p \log \text{rank}(M_f)$ and $p \cdot \log$(the size of the largest fooling set of f) are lower bounds for the total communication needed in this case.

Multi Party Communication Complexity with Related Inputs

[CFL 83] assumed that k processors $P_0, \ldots, P_{k-1}$ communicating over a broadcast channel in order to determine $f(X_0, \ldots, X_{k-1})$ where, initially, P_i knows only X_i, but the X_is can be related. They considered the case where $A_0, \ldots, A_{k-1}$ are arbitrary random integers $\in \{0, \ldots, n-1\}$, each P_i knows $A_0, \ldots, A_{i-1}, A_{i+1}, \ldots, A_{k-1}$ (that is, all the random integers except A_i), and they try to determine whether $A_0 + \cdots + A_{k-1} = n$. They showed that the number of transmitted bits required is related to Ramsey numbers thereby showing, for example, that for 3 participants, the total communication is $O(\sqrt{\log n})$.

Average Complexity for Several Users

[BOH 86] considered N communicators, each having one random bit X_i, communicating over a broadcast channel to determine $f(X_1, \ldots, X_N)$. With the model thus formulated, each protocol corresponds to a decision tree for f. Using the n-dimensional Fourier transform, they obtained a universal lower-bound technique for the average number of bits that must be transmitted.

Almost Unbiased Bits From Two Biased Sources of Randomness

A δ-source is an infinite sequence $X_1, X_2, \ldots$ of random bits. The bits can depend on each other in an arbitrary manner, but for all i and all possible values of $X_1, \ldots, X_i$, the conditional probability that $X_{i+1} = 1$, must be between $\frac{1}{2} - \delta$ and $\frac{1}{2} + \delta$.

The following problem was considered in [SV 84]: given several independent δ-sources and $\epsilon > 0$, find a function f of the bits output by the sources which ensures that $\frac{1}{2} - \epsilon < p(\{f = 1\}) < \frac{1}{2} + \epsilon$. The problem is related to communication complexity because it uses a very similar proof.

[SV 84] showed that for small enough ϵ, no function will do when there is only one source, however, using enough sources, arbitrary small ϵ can be achieved. [Va 85] showed that the inner product function can be applied to just two independent sources to get the desired result. [CG 85] improved on this result by showing that even under less restrictive assumptions, most functions will do.

This problem amounts to finding the largest rectangle in M_f with proportion of 1's deviating from $\frac{1}{2}$ by more than ϵ. This was the essence of the proof of Lemma 5.

Communication with Secrecy Constraints

Let $\rho_f^1(R) \stackrel{\text{def}}{=} |\{(x,y) \in R : f(x,y) = 1\}|/|R|$ be the proportion of inputs in R for which f is 1. In the last topic, the problem of using a function f to produce ϵ-biased bits from two biased sources was related to the size of the largest rectangle R in $\{0,..,n-1\} \times \{0,..,n-1\}$ with $|\rho_f^1(R) - \frac{1}{2}| > \epsilon$. Let $rs(f)$, *the rectangle size of* f, be the smallest integer such that $\{0,..,n-1\} \times \{0,..,n-1\}$ can be partitioned into rectangles of size less than that integer, each with $|\rho_f^1(R) - \rho_f^1(\{0,..,n-1\} \times \{0,..,n-1\})| < \epsilon$. [OE 84] showed that $rs(f)$ is related to the following problem: X and Y are uniformly distributed over $\{0,..,n-1\} \times \{0,..,n-1\}$. $P_{\mathcal{X}}$ knows X, $P_{\mathcal{Y}}$ knows Y and they want to exchange the random variables. However, an eavesdropper who knows the (uniform) distribution of X and Y and the protocol used by $P_{\mathcal{X}}$ and $P_{\mathcal{Y}}$ is trying to update his a priori distribution of $f(X,Y)$. To thwart his efforts, $P_{\mathcal{X}}$ and $P_{\mathcal{Y}}$ have the option of transmitting each bit over a (costly) secure channel (equivalently, they can encrypt that bit). How many bits do they have to send securely in order to keep the eavesdropper's a priori and a posteriori estimates that $f(X,Y) = 1$ ϵ-close?

It was shown that for all functions, only $\log \frac{1}{\epsilon}$ bits must be sent securely. If, in addition, $P_{\mathcal{X}}$ and $P_{\mathcal{Y}}$ can use coin flips to determine their transmissions, then , for all functions f and all probability distributions for (X,Y) (not only uniform), two secure bits are enough to ensure that the eavesdropper will not be able to update his a priori probability estimate that $f(X,Y) = 1$ at all (even when $P_{\mathcal{X}}$ and $P_{\mathcal{Y}}$ don't share the random source).

References

[Ab 78] H. Abelson, 'Lower Bounds on Information Transfer in Distributed Computations', *Proc. of the 19th Annual Symposium on Foundations of Computer Science*, 1978.

[AC 81] R. Ahlswede and I. Csiszár, 'To Get a Bit of Information May be as Hard as to Get Full Information', *IEEE Transactions on Information Theory*, Volume 27, No. 4, July 1981.

[AUY 83] A.V. Aho, J.D. Ullman, and M. Yanakakis, 'On notions of information transfer in VLSI circuits', *Proc. of the 15th Annual ACM Symposium on the Theory of Computing*, April 1983.

[AFR 85] N. Alon, P. Frankl, and V. Rödl, 'Geometrical Realization of Set Systems and Probabilistic Communication Complexity', *Proc. of the 26th Annual Symposium on Foundations of Computer Science*, October 1985.

[BFS 86] L. Babai, P. Frankl, and J. Simon, 'BPP and the Polynomial Time Hierarchy in Communication Complexity', *University of Chicago Technical Report 86-003*, February 1986.

[BOH 86] Y. Brandman, A. Orlitsky, and J. Hennessy, 'A Universal Lower-Bound Technique for Decision Trees and Two-Level AND/OR Implementations of Boolean Functions', *to appear.*

[CFL 83] A. Chandra, M. Furst, and R. Lipton, 'Multi-Party Protocols', *Proc. of the 15th Annual ACM Symposium on the Theory of Computing*, April 1983.

[CG 85] B. Chor, and O. Goldreich, 'Unbiased Bits from Sources of Weak Randomness and Probabilistic Communication Complexity', *Proc. of the 26th Annual Symposium on Foundations of Computer Science*, October 1985.

[DGS 84] P. Duris, Z. Galil, and G. Schnitger, 'Lower Bounds on Communication Complexity', *Proc. of the 16th Annual ACM Symposium on the Theory of Computing*, 1984.

[ElG 83] A. El Gamal, 'A Simple proof of the Ahlswede-Csiszár One-Bit Theorem' *IEEE Transactions on Information Theory*, Volume 29, No. 6, November 1983.

[EO 84] A. El Gamal and A. Orlitsky, 'Interactive Data Compression', *Proc. of the 25th Annual Symposium on Foundations of Computer Science*, October 1984.

[ES 74] P. Erdös and J. Spencer, *Probabilistic Methods in Combinatorics*, Academic Press, 1974.

[Fr 77] R. Freivalds, 'Probabilistic machines can use less running time', Information Processing, IFIP, North Holland, 1977, 839–842.

[HU 79] J.E. Hopcroft and J.D. Ullman, *Introduction to Automata Theory, Languages, and Computation*, Addison-Wesley Publishing Company, 1984.

[Ja' 85] J. Ja'Ja', 'Identification is Easier than Decoding', *Proc. of the 26th Annual Symposium on Foundations of Computer Science*, October 1985.

[JK 84] J. Ja'Ja' and V.K.P. Kumar, 'Information Transfer in Distributed Computing with applications to VLSI', *JACM*, vol.31, no.1, January 1984.

[JKS 84] J. Ja'Ja', V.K.P. Kumar, and J. Simon, 'Information Transfer Under Different Sets of Protocols', *SIAM Journal of Computing*, Vol. 13 No. 4, November 1984.

[LS 81] R.J. Lipton and R. Sedgewick, 'Lower bounds for VLSI', *Proc. of the 13th Annual ACM Symposium on the Theory of Computing*, April 1981.

[MS 82] K. Mehlhorn and E.M. Schmidt, 'Las Vegas is better then Determinism in VLSI and Distributed Computing', *Proc. of the 14th Annual ACM Symposium on the Theory of Computing*, April 1982.

[OE 84] A. Orlitsky and A. El Gamal, 'Communication with Secrecy Constraints' *Proc. of the 16th Annual ACM Symposium on the Theory of Computing*, April 1984.

[OE 85] A. Orlitsky and A. El Gamal, 'Randomized Communication Complexity', *IEEE International Symposium on Information Theory*, June 1985.

[OE 86] A. Orlitsky and A. El Gamal, 'Average and Randomized Communication Complexity', *Submitted for Publication.*

[PE 84] K.F. Pang and A. El Gamal, 'Communication Complexity of Computing the Hamming Distance', to appear, *SIAM J. of Computing.*

[PpS 82] C.H. Papadimitriou and M. Sipser, 'Communication Complexity', *Proc. of the 14th Annual ACM Symposium on the Theory of Computing*, April 1982.

[PtS 84] H. Paturi and J. Simon, 'Probabilistic Communication Complexity', *Proc. of the 25th Annual Symposium on Foundations of Computer Science*, October 1984.

[RY 79] M.O. Rabin and A. Yao, unpublished manuscript (see [Yao 79]).

[SV 84] M. Santha and U.V. Vazirani, 'Generating Quasi-Random Sequences from Slightly Random Sources', *Proc. 25th Annual Symposium on Foundations of Computer Science*, October 1984.

[Sh 48] C.E. Shannon, 'A Mathematical Theory of Communication', *Bell Systems Technical Journal*, July 1948.

[SW 73] D. Slepian and J.K. Wolf, 'Noiseless Coding of Correlated Information Sources', *IEEE Transactions on Information Theory*, July 1973.

[Ti 84] P. Tiwari, 'Lower Bounds on Communication Complexity in Distributed Computer Networks', *Proc. of the 25th Annual Symposium on Foundations of Computer Science*, October 1984.

[Th 80] C.D. Thompson, 'Area-Time Complexity for VLSI', *Proc. of the 11th Annual ACM Symposium on the Theory of Computing*, 1979.

[Ul 84] J.D. Ullman, *Computational Aspects of VLSI*, Computer Science Press, Rockville, Maryland, 1984.

[Va 85] U.V. Vazirani, 'Towards a Strong Communication Complexity Theory or Generating Quasi-Random Sequences from Two Communicating Slightly-Random

Sources'. *Proc. of the 17th Annual ACM Symposium on the Theory of Computing*, May 1985.

[Yao 79] A.C. Yao, 'Some Complexity Questions Related to Distributive Computing', *Proc. of the 11th Annual ACM Symposium on the Theory of Computing*, 1979.

[Yao 81] A.C. Yao, 'The Entropic Limitations on VLSI Computation, *Proc. of the 13th Annual ACM Symposium on the Theory of Computing*, 1981.

[Yao 83] A.C. Yao, 'Lower Bounds by Probabilistic Arguments', *Proceedings of the 24th Annual Symposium on Foundations of Computer Science*, November 1983.

CHAPTER III

Introduction To Information-Based Complexity

J. F. Traub

Computer Science Department
Columbia University
New York, NY 10027

Information-based complexity is based on three assumptions: information is partial, contaminated, and it costs. Its goal is to create a general theory about problems with such information, and to apply the results to solving specific problems in varied disciplines. Some of the questions asked in information-based complexity are listed and some results are indicated. Information-based complexity is contrasted with combinatorial complexity which studies problems such as the traveling salesman problem and linear programming. In combinatorial complexity, information is assumed to be complete, exact, and free. The relation between information theory and information-based complexity is briefly discussed.

Problems with only partial or limited information arise in many disciplines: in economics, computer science, physics, control theory, signal processing, prediction and estimation, scientific and engineering computation, medical imaging, geophysics, decision theory, and artificial intelligence. The goal of information-based complexity is to create a general theory about problems with partial or contaminated information, and to apply the results to solving specific problems in varied disciplines.

Traub and Woźniakowski (1980) and Traub, Wasilkowski, and Woźniakowski (1983) are research monographs on information-based complexity. Expositions and surveys may be found in Traub and Woźniakowski (1984a), Traub (1985), Woźniakowski (1985), Wasilkowski (1985), and Traub and Woźniakowski (1986a).

I'll review what is meant by computational complexity. By the computational complexity of a problem we mean its intrinsic difficulty as measured by the time, space, or other quantity *required* for its solution. For simplicity, I'll confine myself here to time complexity. Equivalently, the computational complexity of a problem is the cost of the optimal algorithm for its solution. Thus computational complexity defines optimal algorithm. For brevity I will usually refer to computational complexity simply as complexity.

Complexity is an invariant of a problem. It is independent of the algorithm, but may depend on the model of computation. See Traub (1985) for further discussion.

In general, determining the complexity of a problem is difficult. One establishes a lower bound by proving that a faster way of performing a task cannot exist; and an upper bound, which is the cost of a particular algorithm. The actual complexity is caught from above and below by these bounds. The deeper question is the lower bound, but good upper bounds are also very important.

There is sometimes confusion because people use the word complexity when they refer to the cost of an algorithm. When I say complexity I'll always mean complexity of the problem.

Basic concepts of information-based complexity include *information*, *partial information*, *contaminated information*, and *uncertainty*. I'll use an example drawn from modeling of the human visual system to illustrate these concepts.

How are humans able to see the world? The late David Marr of M.I.T. and his colleagues have developed a computational model of the human visual system, Marr and Poggio (1977), Marr (1981), Grimson (1981). I am going to give a simplified description of a small portion of this model. I want to point out, parenthetically, that similar issues arise when we design a vision system for a robot.

Imagine you're looking at an automobile. You can see what shape it has because, roughly speaking, your brain has performed a number of processes at various stages upon images it has received. For example, at one stage it has outlined the images of the various surfaces of the car by detecting the edges that separate them, such as the edge that separates the image of the windshield from the image of the hood. You can detect this edge because there's a sudden change in the slope of the surface; the window and hood do not join *smoothly*.

In the next stage, the human visual system identifies the three-dimensional shapes of the various surfaces. This stage will serve as our example.

How do you infer the shape of the hood? A depth value is the subjective distance to a point on the object as perceived by the the viewer. The model assumes that by binocularity or other means you obtain a *finite*, number of depth values. In general, between any pair of depth values the hood could have any shape. However, the visual system uses the assumption that the hood is smooth and therefore cannot change too much between depth values. (This notion of smoothness can be made mathematically precise.) Knowing the finite number of depth values and the smoothness of the surface, the visual system *approximately* determines the shape of the hood.

I'll use this example of determining the shape of the hood to introduce some fundamental concepts.

The first concept is *information*. I want to emphasize we do not mean information in the sense of Shannon, Shannon (1948). As you know, Shannon information is a certain number, the *information content* of a message, not the "information" contained in the message. Near the end of this paper I will briefly discuss connections between information theory and information-based complexity.

For present purposes, information is what is known about the problem to be solved. In determining the shape of the hood, the information is the finite number of depth values and the assumed smoothness of the surface. Because we may want to regard the surface smoothness as fixed, and study the effect of varying the depth values, we often regard the set of depth values as the information.

The number of depth values is finite. Many different surfaces may have the same depth values; there are not enough depth values to uniquely determine the surface. We say the information is limited, or *partial*.

Furthermore, the subjective distance perceived by the viewer is only an estimate of the true distance. Thus the information is *contaminated* by error.

Because the information is partial and contaminated, the problem of determining the shape of the hood can be solved only approximately. Alternatively, we can say there must be *certainty* in the answer and this uncertainty is inherently caused by the available information. It should be clear that partial or contaminated information always leads to inherent uncertainty.

As a second example, I'll use a mathematical problem. It is a simple problem, the computation of a definite integral. For most integrands we cannot compute the integral utilizing the fundamental theorem of the calculus since the antiderivative is not a "simple" function. We have to approximate the integral numerically. Usually, the integrand is evaluated at a finite number of points. The information is the values of the integrand at these points. Since an infinite number of integrands have the same values at these points, the information is partial. The integral is estimated by combining the integrand values. In addition, there will be round-off error in evaluating the integrand, and therefore the information is contaminated. Since with the information we're using we don't know the integrand, there is intrinsic uncertainty in the answer.

This example differs from the previous one in that we started with complete and exact information. The integrand was specified exactly as a function. But we couldn't use that information to solve our problem. We had to throw away our complete and exact information and replace it by partial and contaminated information.

Problems with partial and contaminated information arise elsewhere. In medical imaging, such as a CAT scan, we approximate the shape of internal organs from certain density measurements. In geophysical exploration, we estimate the location of mineral deposits or oil reservoirs by measurements taken at the earth's surface, Twomey(1977).

Two of the basic assumptions of information-based complexity are that information is partial and contaminated. There is one further assumption — information costs. For example, in mineral exploration, a seismologist might set off explosions whose effect is measured by sensors. That's an expensive process. THESE THREE ASSUMPTIONS ARE FUNDAMENTAL: INFORMATION IS PARTIAL, INFORMATION IS CONTAMINATED, AND INFORMATION COSTS.

I'll contrast that with the usual assumptions concerning the traveling salesman problem. The available information specifies just one task. That is, all the intercity distances are given; therefore the information is *complete*. Furthermore, the intercity distances are assumed to be given without error. Therefore the information is *exact*. Finally, the intercity distances are provided without charge. Thus the information is *free*. For the traveling salesman problem then, information is assumed to be complete, exact, and free. These assumptions are also made for other important problems. For example, for linear programming the information is the matrix and linear cost function.

Computational complexity is the study of the minimal cost for solving a problem. Computational complexity comes in two flavors, depending on which set of assumptions are made about information. Is information complete, exact, and free, as with the traveling salesman problem? We call this combinatorial complexity. Or is the information partial, contaminated, and priced, as with a host of other real-world problems?

What gives information-based complexity its particular flavor is based on the observation that the quality and amount of information determine the irreducible uncertainty in the solution. This irreducible uncertainty does not depend on any notion of algorithm. This observation enables us to decompose the problem of determining the computational complexity of solving a problem into an information phase and a combinatory phase. It's a very powerful principle.

Thus information-based complexity has both an *informational level* and a *combinatorial level.* At the information level we answer questions such as:

- What is the intrinsic uncertainty in the solution of a problem due to the available information?
- How much information is needed to limit the uncertainty to a specified level? For example, in geophysical exploration, how many measurements must the seismologist make?
- What is the optimal information? Again, using our geophysical exploration
- example, where on the earth's surface should the
- seismologist place his instruments?
- Can the information be decomposed for parallel or distributed computation?

At the combinatorial level we answer questions such as:

- What is the minimal cost of combining the information to obtain an answer?

The central question of information-based complexity is the following:

- What is the computational complexity of solving a problem for a given level of uncertainty?

Answering this question requires both the information and combinatorial levels.

The answers to these questions depend on the setting. In both combinatorial complexity and information-based complexity we judge algorithms by their performance in various settings including worst case, average case, and probabilistic. I want to indicate some of the settings studied in information-based complexity.

In a *worst case* setting we want to minimize the worst case cost while guaranteeing that the error for the most difficult problem instance does not exceed a preassigned threshold. This is sometimes called a minimax criterion.

In an *average case* setting we want to minimize the expected cost while guaranteeing that the average error does not exceed a preassigned threshold.

In a *probabilistic* setting we want to minimize the expected cost while requiring that the probability of a large error is small. If a problem has a yes or no answer, then we require that the probability of a wrong answer is small.

Other settings are also of interest in information-based complexity. In a study of clock synchronization in distributed networks Wasilkowski (1985) uses a worst case setting in which he averages over the delays. In a study of whether adaptive information is more powerful than nonadaptive information for stochastic information, Kadane, Wasilkowski, and Woźniakowski (1984) minimize the worst case cost, guarantee the expected error, and average over the information error which is assumed to be stochastic.

In another setting of interest we want to minimize the expected cost for the most difficult problem instance. Such a setting is of interest for problems such as aircraft design and structural design for earthquakes.

A discussion of these and other settings may be found in Traub and Woźniakowski (1986a).

At the beginning of this paper I said that the goal of information-based complexity is to create a general theory about problems with partial or contaminated information, and to apply the results to solving specific problems in varied disciplines. I now will talk about several general results and then briefly indicate applications.

I'll discuss results concerning questions one and four at the information level.

Recall that the first question was:

- What is the intrinsic uncertainty in the solution of a problem due to the available information? There exists a quantity which measures this intrinsic uncertainty, Traub, Wasilkowski and Woźniakowski, (1983). The radius of information is fundamental, not only to answering the first question but also to answering some of the other questions I posed. To emphasize this I will state the information principle: *There exists a quantity called the radius of information which measures the intrinsic uncertainty of solving a problem for specified information.*

This radius of information exists in every setting we study. It exists in the worst case, average case, and probabilistic settings.

In principle, one can always compute the radius of information for any problem and any specified information. For a particular instance, it may be technically very difficult to do this. Of course, we're used to that. For example, in principle one need only follow some basic prescriptions of quantum mechanics to solve certain problems of interest to physicists and chemists, but it's quite a different matter to actually follow the prescription, either symbolically or numerically, for a particular instance.

Next I'll discuss question four. The fourth question is:

– Can the information be decomposed for parallel or distributed computation?

One important way to solve a problem on a parallel or distributed system is to decompose the problem. That will be a subject for investigation for as far ahead as I can see. In particular, can the task of obtaining information be decomposed? That's the motivation for the above question.

I'll use a simple example to introduce the notions of sequential (adaptive) and parallel (nonadaptive) information. Consider estimating the value of an integral using as the information, values of the integrand at a finite number of points.

Imagine I've evaluated the integrand at a number of points and therefore have some idea of the shape of the curve. Can I take advantage of that knowledge to choose the next point for evaluation? For example, should I evaluate the integrand where its slope is large, that is, where its shape is changing rapidly?

The strategy I've just described uses sequential information. It is sometimes called *adaptive* information. *Parallel* information, on the contrary, means that I perform all the evaluations simultaneously without having any opportunity to take advantage of what we learn about the shape of the curve.

Those are informal descriptions of sequential and parallel information. Precise descriptions may be found in the literature.

The desirable results for distributed or parallel computation is that sequential information is not more powerful than parallel information; that adaption does not help. To see this for the integration example, assume I have p processors and that it takes a processor one unit of time to evaluate a function at a point. Then n functions evaluated sequentially cost n units of time, whereas n functions evaluated simultaneously, that is, in parallel, cost $\lceil n/p \rceil$ units of time. Thus the cost of collecting information is reduced by a factor of p, which is linear speed-up. Since the cost of evaluating the integrand dominates the cost of combining evaluations to produce an answer, the speed-up in computing an approximation to the integral is close to linear in the number of processors. See Traub and Woźniakowski (1984a, Section 4.5) for further discussion of speed-up.

The above discussion is predicated on the assumption that adaption does not help. However, intuition suggests that adaption does help.

The question of whether adaption helps was first analyzed for a worst case model. If a problem is linear, adaption does not help. Examples of linear problems are linear functionals such as intergration and differentiation, approximation, signal recovery, and linear partial differential equations. The vision example I described earlier is a special case of approximation. See Traub and Woźniakowski (1980), Traub, Wasilkowski, and Woźniakowski (1983).

We believed, along with others, that the result that adaption doesn't help, was due to the worst case model. To our surprise we were able to show that adaption does not help on the average. This was established in great generality. It is enough to assume that the problem is specified by a linear operator and that the distribution of problem instances is governed by a measure with a certain symmetry property, such as a Gaussian or Wiener measure. See Traub, Wasilkowski, and Woźniakowski (1984a), Wasilkowski and Woźniakowski (1984), Lee and Wasilkowski (1985), and Wasilkowski (1985). What if the information is contaminated with stochastic error? Kadane, Wasilkowski, and Woźniakowski (1984) recently showed that again adaption does not help, provided only that the information noise is unbiased.

In all the work cited above the problem is specified by a linear operator. Perhaps adaption helps generally for nonlinear problems. However, Sikorski (1984) has shown that adaption does not help for the approximate solution of nonlinear equations satisfying a Lipschitz condition.

I must emphasize that there are, of course, many problems for which adaption helps. (See Traub, Wasilkowski, and Woźniakowski (1984b) for an especially constructed example. We are currently exploring problems, arising in practice, where adaption helps.) We *have*, however, identified many important problems for which we can prove that it does not and for which the answer to the question "Can the information be decomposed for parallel or distributed computation?" is yes.

I want to turn to applications of information-based complexity beginning with a general discussion of distributed systems.

There are two reasons for distributed systems. The first is that although a centralized system could be used, we select a distributed system for the sake of, say, efficiency. The second reason is that the problem is inherently distributed; examples include resource allocation in a decentralized economy, traffic networks, and reservation systems.

Consider now a large distributed system. One possibility is that the total system has complete information but the nodes have only local information, say, about themselves and their neighbors. Thus the information is distributed over the system. To give the nodes information about the total system would cost too much in time and/or space. Thus, decisions are made at nodes which have only partial information and that means a solution with uncertainty. In a dynamic system, even if complete information is initially available at the nodes, we cannot afford to update that information over time.

So far, I've assumed the total system has complete information. Of course, often even the total system has only partial or contaminated information and what I've said holds in spades.

An application of this discussion is that even the problems that are now exactly solved on a uniprocessor will be only approximately solved in the distributed environments of the future.

Let me return to the present. If a user wished to apply information-based complexity he provides just three items. They are:

- 1. A mathematical formulation of his problem and the setting (say, average case) in which he's interested,
- 2. The available information,
- 3. The model of computation; that is what information operations and combinatory operations are permitted and how much they cost.

That's all the user provides. The theory then provides everything else. For examples, the theory tells the user such things as how much information is needed to limit the uncertainty to a specified level, whether the information can de decomposed for parallel or distributed computation, the optimal algorithm, and the computational complexity of the problem.

At the beginning of this article I indicated some of the disciplines for which only partial or limited information is available and to which information-based complexity can therefore be applied. We indicate some of the application areas which are currently being pursued. For each application we list one recent paper or book, often a survey. An extensive discussion of applications may be found in Traub and Woźniakowski (1986a).

Vision, Medical Imaging, Optical Flow

Lee (1985)

Synchronization of Clocks in Distributed Systems

Wasilkowski (1985)

Nonlinear Constrained Optimization

Nemirovsky and Yudin (1983)

Prediction and Estimation

Milanese, Tempo, and Vicino (1986)

Partial Differential and Integral Equations

Werschulz (1985a)

Ordinary Differential Equations

Kacewicz (1984)

Nonlinear Equations

Sikorski (1985)

Integration

Traub and Woźniakowski (1980, Section 6.4)

Topological Degree

Boult and Sikorski (1986)

Ill-Posed Problems

Werschulz (1985b)

Large Eigenpair Systems

Kuczyński (1986)

Large Linear Systems

Traub and Woźniakowski (1984b).

A natural question concerns connections between information-based complexity and information theory. The concepts of information and entropy as developed in thermodynamics by Boltzmann and Clausius and in information theory by Shannon are among the most significant and fruitful scientific ideas of the last hundred years. It seems to me, however, that the kind of questions that were posed earlier are not addressed by information theory. Indeed, as indicated above, information in the sense of Shannon is a certain number, the information content of a message, not the "information" contained in the message. It turns out, however, that there are connections between certain notions of entropy and information-based complexity. For example, Traub and Woźniakowski (1980, Section 7.4) have reported on the relation between Kolmogorov's notion of ε-entropy and the second question at the information level on how much information is needed to limit the uncertainty to a specified level.

More generally, uncertainty and information are basic concerns of a number of disciplines including information theory, statistics, economics, and control. Various measures of uncertainty and information have been proposed. For discussions see, for example, Brillouin (1956), Fisher (1950), Gallager (1968), Kullback (1961), Marschak and Radner (1972), Pinsker (1964), Shannon (1948), Theil (1967), and Wiener (1948).

In information-based complexity the radius of information measures the uncertainty. In Traub and Woźniakowski (1986b) we compare and contrast various measures of uncertainty and information including Shannon entropy, Kullback-Leibler entropy, Fisher information, variance, and radius of information.

The aim of information-based complexity is to find unity across a great variety of what seem, at first, to be totally different disciplines. I believe we have made a beginning.

References

[1] Boult, T. and Sikorski, K. (1986), Complexity of Computing Topological Degree of Lipschitz Functions in n Dimensions, to appear in *Journal of Complexity*, **2**.

[2] Brillouin, L. (1956), *Science and Information Theory*, Academic Press, New York.

[3] Fisher, R. A. (1950), *Contributions to Mathematical Statistics*, John Wiley and Sons, New York.

[4] Gallager, R. G. (1968), *Information Theory and Reliable Communication*, John Wiley and Sons, New York.

[5] Grimson, W. E. L. (1981), *From Images to Surfaces: A Computational Study of the Human Early Visual System*, MIT Press, Cambridge, Mass.

[6] Kacewicz, B. Z. (1984), How to Increase the Order to Get Minimal-Error Algorithms for Systems of ODE, *Numer. Math.*, **45**, 93–104.

[7] Kadane, J. B., G. W. Wasilkowski, and H. Woźniakowski (1984), Can Adaption Help on the Average for Stochastic Information?, *Report*, Columbia University.

[8] Kolmogorov, A. N. and V. M. Tihomirov, ε-Entropy and ε-Capacity of sets in Functional Spaces, *Usp. Mat. Nauk.* **14**, 3–80 (in Russian), (1959). Translation in *Amer. Math. Soc. Transl.* **17**, 277–364, (1961).

[9] Kuczyński, J. (1986), On the Optimal Solution of Large Eigenpair Problems, to appear in *Journal of Complexity*, **2**.

[10] Kullback, S. (1961), *Information Theory and Statistics*, John Wiley and Sons, New York.

[11] Lee, D. (1985), Optimal Algorithms for Image Understanding: Current Status and Future Plans, *Journal of Complexity*, **1**, 138–146.

[12] Lee, D. and G. W. Wasilkowski (1985), Approximation of Linear Functionals on a Banach Space with a Gaussian Measure, *Report*, Columbia University. To appear in *Journal of Complexity*, **2** (1986).

[13] Marr, D. (1981), VISION: *A Computational Investigation in the Human Representation and Processing of Visual Information*, W. H. Freeman, San Francisco.

[14] Marr, D. and T. Poggio (1977), From Understanding Computation to Understanding Neural Circuitry, *Neuroscience Research Program Bulletin* **15**, 470–488.

[15] Marschak, J. and R. Radner (1972), *Economic Theory of Teams*, Yale University Press, New Haven.

[16] Milanese, M., R. Tempo, and A. Vicino (1986), Strongly Optimal Algorithms and Optimal Information in Estimation Problems, to appear in *Journal of Complexity*, **2**.

[17] Nemirovsky, A. S. and D. B. Yudin (1983), *Problem Complexity and Method Efficiency in Optimization*, Wiley-Interscience, New York.

[18] Pinsker, M. S. (1964), *Information and Information Stability of Random Variables and Processes*, Holden-Day, San Francisco.

[19] Shannon, C. E. (1948), Mathematical Theory of Communication, *Bell Syst. Tech. J.* **27**, 379–423, 623–658.

[20] Sikorski, K. (1984), Optimal Solution of Nonlinear Equations Satisfying a Lipschitz Condition, *Numer. Math.* **43**, 225–240.

[21] Sikorski, K. (1985), Optimal Solutions of Nonlinear Equations, *Journal of Complexity*, **1**.

[22] Theil, H. (1967), *Economics and Information Theory*, North-Holland, Amsterdam.

[23] Traub, J. F. (1985), Information, Complexity, and the Sciences, University Lecture, Columbia University.

[24] Traub, J. F., G. W. Wasilkowski, and H. Woźniakowski (1983), *Information, Uncertainty, Complexity*, Addison-Wesley, Reading, Mass.

[25] Traub, J. F., G. W. Wasilkowski, and H. Woźniakowski (1984a), Average Case Optimality for Linear Problems, *Journal Theoret. Comp. Sci.*, **29**, 1–25.

[26] Traub, J. F., G. W. Wasilkowski, and H. Woźniakowski (1984b), When is Nonadaptive Information as Powerful as Adaptive Information?, *Proceedings of the 23rd IEEE Conference on Decision and Control*, 1536–1540.

[27] Traub, J. F. and H. Woźniakowski (1980), *A General Theory of Optimal Algorithms*, Academic Press, New York, NY.

[28] Traub, J. F. and H. Woźniakowski (1984a), Information and Computation, chapter in *Advances in Computers* **23**, M. C. Yovits, editor, Academic Press, New York 35–92.

[29] Traub, J. F. and H. Woźniakowski (1984b), On the Optimal Solution of Large Linear Systems, *Journal of the ACM*, **31**, 545–559.

[30] Traub, J. F. and H. Woźniakowski (1986a), Information-based Complexity, to appear in *Annual Review of Computer Science* **1**, Annual Reviews, Inc., Palo Alto.

[31] Traub, J. F. and H. Woźniakowski (1986b), Measures of Uncertainty and Information, in progress.

[32] Twomey, S. (1977), *Introduction to the Mathematics of Inversion in Remote Sensing and Indirect Measurement*, Developments in Geomathematics **3**, Elsevier, Amsterdam.

[33] Wasilkowski, G. W. (1985), Average Case Optimality, *Journal of Complexity* **1**, 107–117.

[34] Wasilkowski, G. W. (1985), Clock Synchronization Problem with Random Delays, *Report*, Columbia University.

[35] Wasilkowski, G. W. and H. Woźniakowski (1984), Can Adaption Help on Average?, *Numer. Math.*, **44**, 169–190.

[36] Werschulz, A. G. (1985a), Complexity of Differential and Integral Equations, *Journal of Complexity*, **1**.

[37] Werschulz, A. G. (1985b), What is the Complexity of Ill-posed Problems?, *Report*, Columbia University.

[38] Wiener, N. (1948), *Cybernetics*, John Wiley and Sons, New York.

[39] Woźniakowski, H. (1985), A Survery of Information-Based Complexity, *Journal of Complexity* **1**, 11–44.

CHAPTER IV

Complete Representations For Learning From Examples

E. B. Baum

Jet Propulsion Laboratory
California Institute of Technology
Pasadena, CA 91109

Valiant has previously defined what it means for F to be learnable from example by G, where F is a class of concepts and G a class of programs or representations. This chapter emphasizes that whether F is learnable by any class of representations is more interesting than whether F is learnable by a particular class G. The 'transitive closure' of a class of representaions is defined. Feedforward neural nets are the transitive closure of linear threshold functions. Feedforward neural nets and also the transitive closure of 3-DNF formulae are shown to be 'complete' representations in the sense of Cook's Theorem, that is, for any F and any G such that the functions $g \in G$ can be evaluated in polynomial time, if F is learnable by G then F is learnable by feedforward neural nets. Other results include an algorithm for learning unrestricted conjunctive normal forms, answering (in a sense) an open question of Valiant; an algorithm for learning an arbitrary class of functions in exponential time but using only a number of examples bouned, as in Valiant's protocol for learning, by a polynomial in the size of the function to be learned; and a proof that no class containing sufficiently many distinct functions is learnable by any repreentation under a strong but pragmatically interesting time restriction.

Humans seem able to learn many difficult tasks, such as speech recognition, face recognition, or walking. As it becomes clear that explicit programming of computers to duplicate human abilities is an arduous if not impossible task, interest in learning increases. Researchers have studied heuristic algorithms, such as back propagation for learning on feed-forward neural nets[20,23], genetic algorithms[9], and a host of methods for parameter or signature table fitting[6]. It is hoped that these efforts, and neural net methods in particular, will not only lead to practical applications, but to an increased understanding of computation in the brain as well. Both back propagation and genetic algorithms have had considerable experimental success in abstracting rules from real world examples, but there is little rigorous understanding of their abilities, their scaling behavior, their optimality, their results, or indeed when they can be expected to converge[20,9].

About four years ago Valiant embarked on a rigorous program to understand the limitations imposed by complexity theory on learning[22,12,16]. He studied generalization by assuming that 'Nature' provides positive and negative examples according to some simple rule and asked under what conditions and for what classes of rules it is possible to deduce in polynomial time a reasonable classifier of future examples. This promising approach has produced both positive examples of polynomial time learning algorithms for interesting problems and proofs that certain classes of problems can not be learned in polynomial time by certain representations, unless P=NP. Recently it has become clear (both in the heuristic approaches and in Valiant's complexity theory approach) that the representation used to capture the examples is crucial to the success of the learning. In my pragmatic view, the only considerations in choosing a representation for learning are: firstly that it be effective for the problem to be learned, and secondly, that it be as compatible as possible with the available hardware- or in the case of the brain, wetware.

In this paper I will prove that there exist 'complete' representations able to capture any problem learnable in polynomial time by any representation whatsoever. I will give two examples of complete representations- one being feedforward neural networks. This result parallels exactly Cook's Theorem[5,7]. I show that any learning algorithm capturing a particular class of example functions using a given class of internal representations can be mapped into one using as representation the "transitive closure of 3-DNF" (which I define), or alternatively, a class of feedforward neural nets. Having shown this, it is natural to look for classes of problems which are not learnable by feedforward networks and thus simply not learnable from examples. A reasonable possibility is that feedforward neural networks themselves form such an unlearnable class.

Section 2 briefly reviews Valiant's definitions and surveys some of his results. Section 3 contains several remarks regarding these definitions which will be useful in the following, including the observation that if we somehow are able to guess an adequate representation of the function to be learned, we may quickly check this by testing it on examples.

In Section 4 I prove three theorems. Theorem 2 gives an algorithm for learning the class of unrestricted functions in conjunctive normal form (CNF's). This answers in one sense an open question of Valiant[22]. It will be clear, however, that whether an algorithm learns, in polynomial time, the class of CNF's depends on which "reasonable encoding" is used to describe the CNF's. The algorithm given learns the CNF's under an encoding which seems, to me, as "natural" as any. The right message to draw from Theorem 2 may thus be that the question posed is not as central as had perhaps been previously believed.

Theorem 4 shows that no class containing 2^{c^n} distinct functions, for $1 < c \leq 2$, can be learned by any representation in time polynomial in the number n of feature variables describing the world, and polynomial in the accuracy of the learning, but independent of the size of the function to be learned. This time restriction, although stronger than the one allowed in our definition of learning, is nonetheless practically important. This indicates, in my opinion, that for practical considerations attention should be restricted to classes containing only $2^{p(n)}$ functions, for some polynomial p. The practically important question is then which classes of this size are both learnable and useful for describing the world.

Theorem 5 gives an algorithm which can learn *any* class of functions, in exponential time, but accessing only a number of examples bounded, as in our learning protocol, by a polynomial in the size of the function to be learned. A similar result has been presented by Blumer et. al. in [3]. This result is grounds for optimism for eager cryptographers or astronomers who are willing to invest vast quantities of computer time but can gather limited information. It however is grounds for pessimism for theorists who hope to prove classes of functions are unlearnable without recourse to complexity theory assumptions.

Section 5 defines the transitive closure of a representation, and in particular describes the transitive closure of 3-DNF, as well as feedforward neural nets- which are the transitive closure of linear threshold functions. Section 6 gives the main result that the closure of 3-DNF and also feedforward neural nets, as well as the transitive closure of any class of functions sufficiently general that the closure contains *and*, *or*, and *not* gates, form complete representations for learning from examples. Thus neural nets, for example, are universal learners, capable of learning any class of concepts which can be learned by any useful method in feasible time.

§2: Learnability

Valiant[22,12,16] discusses learning in a world where objects are represented by vectors $\vec{x} \in \{0, 1, *\}^n$ which describe their features. The symbol $*$ denotes that a variable is undetermined. A vector is *total* if every variable is determined, i.e. each feature is given as present or absent. A *Boolean Function* f is defined as a mapping of the 2^n possible total feature vectors to $\{0,1\}$. A Boolean function is extended to be a *concept* c by defining $c(\vec{x}) = 1$ if every assignment of 1 or 0 to the undetermined variables in $\vec{x}$ yields a vector $\vec{x}'$ for which $f(\vec{x}') = 1$, and defining $c(\vec{x}) = 0$ otherwise. Vectors $\vec{x}$ for which $c(\vec{x}) = 1$ (0) are called *positive* (respectively *negative*) examples. In this paper I will confine myself for simplicity to total vectors, so that I will not distinguish between Boolean functions and concepts. My results will be seen to trivially extend to the more general case.

A Boolean function f may be represented in many ways- for example as a fortran program P which outputs a 1 or a 0 when given as data any total vector $\vec{x}$. Alternatively a Boolean formula over the feature variables $\vec{x}$ can represent a function. One is interested in algorithms capable of learning from examples whole classes of concepts (as any single concept is trivially 'learnable' by the algorithm which simply guesses that particular concept). Thus Valiant and collaborators discuss classes of *programs* as being learnable, where a program is defined as an algorithmic specification with binary output. This may be described as an explicit procedure, or some knowledge representation with an associated evaluation procedure[16]. Although in Valiant's earlier work[22] the distinction was not made, more recent papers [12,16] have remarked that one's ability to learn a given class of concepts may depend on the representation one chooses to learn with. Let F be a class of concepts and G a class of representations. A class of concepts F will be defined to be learnable with respect to a representation class G when there exists a polynomial time algorithm using only positive and negative examples, that, for any program $f \in F$, can find a program $g \in G$ which adequately represents f. This will be made explicit shortly.

Following the discussion of Garey and Johnson[7], and more generally the Church-Turing thesis[21], I will assume that any program or algorithm can be encoded as a string over some set Σ of symbols, and thus can be viewed as a program for a deterministic Turing machine. I will return to this important point in section 5.

The learning algorithm has access to an oracle called EXAMPLES which produces on request either a negative or a positive example- that is a feature vector $\vec{x}$ for which $f(\vec{x}) = 0$ or 1. The example is generated according to some fixed but unknown probability distribution D^- (D^+). Most of Valiant's work concerns distribution free learning, where nothing is assumed about the distributions D^- and D^+ except that $\sum_{f(\vec{x})=1} D^+(\vec{x}) = 1, \sum_{f(\vec{x})=1} D^-(\vec{x}) = 0, \sum_{f(\vec{x})=0} D^-(\vec{x}) = 1$, and $\sum_{f(\vec{x})=0} D^+(\vec{x}) = 0$. It is also of interest to study learning in the case where D^+ and D^- are suitably restricted, for example some programs may be learnable with respect to almost any (but not every) distribution. My results will extend to this case. A parameter T describes the size of the program $f \in F$ to be learned.

Definition[16]: Let F and G be classes of programs representing concepts. Then *F is learnable from examples by G* if and only if there exists a polynomial p and a (possibly randomized) learning algorithm A that accesses f only via EXAMPLES(f) such that for all $f \in F$, for all D^+, D^-, and for all $\epsilon > 0$, the algorithm A halts in time $p(n, T(f), 1/\epsilon)$ and outputs a program $g \in G$ that with probability at least $1 - \epsilon$ has the properties[1]:

$$\sum_{g(\vec{x})=0} D^+(\vec{x}) < \epsilon$$

and

$$\sum_{g(\vec{x})=1} D^-(\vec{x}) < \epsilon.$$

If F is learnable by F itself, I will describe F as *self-learnable*. Pitt and Valiant[16] describe this as 'learnable'. I will define instead F to be *learnable* if there exists *any* representation G such that F is learnable from examples by G. In my opinion, a pragmatic learner should be willing to use any representation in order to discover the underlying regularities in the world and predict their behavior. Moreover, in the real world, learners are not generally given a class F from which examples are to be drawn- knowledge of this representation might *a priori* be expected to be powerful information which it is unfair to assume. A sense in which Valiant's approach escapes this objection will be clarified at the end of the next section, in the discussion leading to Theorem 1.

Valiant's definition of learning is very well motivated by a model of Nature as providing examples of a concept to be learned according to some unknown probability distribution. One can not reasonably demand a learning algorithm that correctly classifies all examples, as some cases may occur with very low probability and thus one might have to examine a huge number of examples to learn this. Thus g is allowed to be wrong on a fraction ϵ of the examples. At the same time, it is possible that Nature will roll the dice badly and present us with an uncharacteristic sequence of examples. Thus one should be allowed a small probability (say ϵ) of producing a highly inaccurate program. Valiant has shown that certain interesting classes of programs can be learned in a distribution independent fashion using a simple probabilistic result[22].

Valiant considered several classes of concepts including CNF, k-clause CNF, and k-CNF formulae; and DNF, k-term DNF, and k-DNF formulae. A CNF formula is a Boolean formula in conjunctive normal form, that is a product of clauses each of which is a sum of literals. (A literal is a Boolean variable u or its negation $\bar{u}$.) For example: $(x_1 + \bar{x}_3)(x_2 + \bar{x}_1 + \bar{x}_4)$. k-CNF is the class of CNF formula with at most k literals in each clause. k-clause CNF is the class of CNF formulae with at most k clauses. A DNF formula is a Boolean

[1] g must of course also halt and output 0 or 1 on every input vector $\vec{x}$.

formula in disjunctive normal form, that is a sum of terms each of which is a product of literals, for example $x_1\bar{x}_3\bar{x}_4 + \bar{x}_2\bar{x}_3$. A k-DNF formula is a DNF formula with at most k literals in each term. k-term DNF is the class of DNF formulae with at most k terms.

Valiant and his collaborators[22,16,12] have proved results including: for all $k \geq 2$, k-term DNF is not self learnable; for all $k \geq 1$ k-CNF is self learnable; k-term DNF is learnable by k-CNF; and thus k-term DNF$\bigcup$ k-CNF is self learnable.

I will conclude this section by briefly reviewing Valiant's self-learning algorithm for k-CNF expressions, as I believe this algorithm is vital in understanding this approach to the complexity of learning. Valiant[22] first defines $L(\epsilon, S)$, for any real number: $0 < \epsilon < 1$ and any positive integer S, as the smallest integer such that in $L(\epsilon, S)$ independent Bernoulli trials each with probability at least ϵ of success, the probability of having fewer than S successes is less than ϵ. Valiant proves that $L(\epsilon, S) \leq (2/\epsilon)(S - log_e(\epsilon))$.

Now one may learn a concept represented by a k-CNF formula f by the following algorithm. Start with an initial formula g which is the product of all possible clauses of up to k literals. There are thus fewer than $(2n)^{k+1}$ clauses in g. Call EXAMPLES $L(\epsilon, (2n)^{k+1})$ times to obtain positive examples of f. For each vector x so obtained, delete all clauses in g not containing a literal determined to be true in x. Valiant shows the set $\{x : x \Rightarrow g\}$ is always a subset of $\{x : x \Rightarrow f\}$. This is clear from the definition of g and implies that g never misclassifies a negative example. Valiant next shows that with probability $1 - \epsilon$, after $L(\epsilon, (2n)^{k+1})$ examples, a g_o will have been produced which has probability less than ϵ of misclassifying a positive example drawn from the distribution. This follows from the definition of L. Let $X = \sum_{g_o(x)=0} D^+(x)$. If $X \geq \epsilon$, then $L(\epsilon, (2n)^{k+1})$ Bernoulli experiments (i.e. calls of EXAMPLES), each with probability at least ϵ of success (i.e. finding an x such that $x \not\Rightarrow g$) have been performed without obtaining $(2n)^{k+1}$ successes (since manifestly not all clauses have been removed from g). By definition of L, the probability of this event occuring is less than ϵ.

§3: Remarks on Learnability

This section contains several remarks and clarifications concerning the definitions given in section 2. In particular I observe in Theorem 1, that if we somehow guess an adequate representation of the function f to be learned, we may quickly check this fact. This observation will be applied in proving Theorems 2 and 5 in section 4. I also define the notion of a p-time representation, which will be necessary in the statement of Theorem 6.

Notice first that Pitt and Valiant's[16] definition does not demand that the value of g be computable in time polynomial either in T or n. For practical purposes one would like to derive a representation which allows prediction of whether a given example is positive or negative. If for all $g \in G$ and for all $\vec{x}$, $g(\vec{x})$ is computable in time polynomial in n and T(g), then I will call G a *p-time representation.* Note that the size T(g) of an expression produced by a learning algorithm in time polynomial in n, ϵ, and T(f) will itself be polynomial in these quantities. It is difficult to see how a learning algorithm A could effectively utilize in polynomial time expressions g which can not even be evaluated in polynomial time. Thus this definition seems to be purely technical, and inclusive of all interesting cases.

Along the same lines, notice that in Valiant's definition, no mention is made of how *long* it might take to compute the function f. For example, a feature vector $\vec{x}$ might encode an instance of the Travelling Salesman Problem, and $f(\vec{x})$ might be 1 if there is a tour less than distance $d(\vec{x})$ and 0 otherwise. Even if f can not be computed in time polynomial in n, it may still be possible to learn it with a p-time representation because the learned algorithm g is allowed a fraction ϵ of errors. Many NP-complete problems, taken with certain distributions of examples, can be solved by *expected*-polynomial time algorithms- or alternatively polynomial time heuristics which are almost always right. It is an interesting question whether any (or indeed every) such problem can be learned from examples *in a distribution independent manner.*

A learning algorithm which is able to learn from examples any function in a given class F of functions uses not only the explicit information encoded in the examples, but also the implicit information that the function to be learned (using whatever representation) is in fact in the class F. This implicit information is of course powerful, and one might worry to what extent a real world learner, with no *a priori* certainty about the class F of functions, might benefit from the formalism developed by Valiant and coworkers. Learning algorithms for given classes of functions escape this philisophical dilemma in a certain sense. An algorithm A to learn any function in a class F may be applied so that, in any instance, if the function f to be learned is in F, A will (with probability $1-\epsilon$) learn f; and if f is not in F, but A announces that it has learned f anyway, then the user can have confidence $1-\epsilon$ that the representation g produced by A will be able to classify future examples at least a fraction $1-\epsilon$ of the time. Thus the user may simply guess at the class of functions to be learned with confidence that, (a)if his guess is correct, his algorithm will succeed, and (b) independent of his confidence in his guess, if his algorithm seems to succeed, then he has learned the given function. (The interesting question of learning what the class F is given exemplar functions $f_i \in$F will not be addressed here.)

For example, the algorithm Valiant gives for learning k-CNF formulae[22], relies only on positive examples, and has the property that it generates a k-CNF formula g which correctly classifies all examples seen in learning and (with probability $1-\epsilon$) will classify correctly (with probability $1-\epsilon$) all future positive examples, assuming the function f to be learned is itself a k-CNF formula. Valiant's proof in fact shows more, namely that even if f is not a k-CNF formula, then (with probability $1-\epsilon$), g, assuming only that it has correctly classified all examples seen in learning, will continue to correctly (with probability $1-\epsilon$) classify positive examples. If f is not a k-CNF, however, g could missclassify every negative example. For example let the algorithm be applied to learn as a 2-CNF the function

$$f = (\bar{x}_1 + \bar{x}_2 + x_3)$$

with $D_-(1,1,0) = 1, D_+$ uniform over (1,0,0), (1,0,1),(0,1,1),(0,1,0), and (1,1,1). The algorithm will yield (incorrectly) $g = (x_1 + x_2)$.

The algorithm may be trivially extended so that, it is not only able to learn any k-CNF formula, but when applied to any set of positive and negative examples generated by any concept of any type, if it is able to explain all the examples generated in learning, then it has probability at least $1-\epsilon$ of correctly classifying all but a fraction ϵ of future examples. To extend the algorithm, after $L(\epsilon,(2n)^{k+1})$ calls of positive examples have been used in Valiant's algorithm, $L(\epsilon,1)$ negative calls may be made, and the k-CNF formula generated in the positive learning phase may be rejected as unable to learn the function if it missclassifies any of the negative examples. The expression g generated by Valiant's algorithm will correctly classify all the negative examples, if f is in fact a k-CNF, and if g correctly classifies all negative examples, the user will have confidence $1-\epsilon$ that any future example is correctly classified, indepedent of any knowledge of the form of f.

More generally, for any problem, if we somehow generate an expression which is successful at predicting examples, we may, with polynomially many calls of EXAMPLES verify this with confidence $1-\epsilon$. Indeed I may easily prove:

Theorem 1: Given expression g, if after $(12/\epsilon)log_e(1/\epsilon)$ calls of positive EXAMPLES, we have made fewer than $12log_e(1/\epsilon)$ mistakes, we may conclude with confidence $1-\epsilon^3$ that p, the probability of error on prediction of positive examples, is less than 2ϵ, and if we make more than $12log_e(1/\epsilon)$ errors, we may conclude with confidence $1-\epsilon^2$, that $p \geq \epsilon/2$; and similarly for negative EXAMPLES.

Proof: For $0 \leq p \leq 1$ and m a positive integer, define $LE(p,m,r)$ as the probability of at most r successes in m independent Bernoulli trials, each with probability of success p, and $GE(p,m,r)$ the probability of at least r successes. Then Kearns et al.[12] quote two facts:
Fact 1. $LE(p,m,(1-\beta)mp) \leq e^{-\beta^2 mp/2}$ and
Fact 2. $GE(p,m,(1+\beta)mp) \leq e^{-\beta^2 mp/3}$, for $0 \leq \beta \leq 1$.
Fact 1 with $\beta = 1/2$ gives the first half of the theorem and Fact 2 with $\beta = 1$ gives the second half. Here, calling an example and checking whether g correctly classifies it constitutes one trial. **Q.E.D.**

In real world applications of heuristic learning algorithms, testing the generated recognition algorithm (sometimes called in this context an "expert system"[6,9,18]) is vital and a sample of the data is generally reserved as a "test set". As acquiring data is frequently a major limiting factor, it is important to use as much of the data as possible for training, yet confidence in the "expert system" is essential. Theorem 1 is thus practically important in giving a rigorous measure of the confidence obtainable from a small test set.

§4: Learnability and Information

This section presents three theorems. I use an improved version of the kind of confidence checking described in Theorem 1 in extending Valiant's algorithm for learning k-CNF's to an algorithm for learning unrestricted CNF's. I also use a similar confidence checking argument in proving that any class of functions can be learned by an algorithm which uses exponential time but accesses only polynomially many examples. The dependence of the learning algorithms on the size $T(f)$ of the function f to be learned is crucial to both of these results and is discussed in detail. I show that no sufficiently large classes of functions can be learned if the polynomial number of examples called is not allowed to depend on $T(f)$. In practical applications, this may be a crucial limitation.

Valiant has given a self-learning algorithm for k-CNF expressions, and asked whether there is any learning algorithm for unrestricted CNF expressions[22, p1138; 12, p292]. To answer this question, one must remember that there are 2^{3^n} possible CNF expressions, and T can be $O(3^n)$. The algorithm given by Valiant for k-CNF expressions can be modified to learn CNF expressions in time polynomial in T(f), although not time polynomial in n.

To see this requires a precise definition of the size of a CNF expression. Order the CNF expressions lexicographically so that any k-CNF expression has smaller order than any CNF expression containing a clause with $k+1$ literals. The size of a CNF expression may be taken as the logarithm of its place on this lexicographic list.

Other orderings of the CNF's are of course possible. For example the CNF's are frequently considered to be ordered by the number of literals in the expression. The algorithm I give will not apply to learning CNF's (in polynomial time) if they are considered in some other ordering than the one described in the preceeding paragraph. (I will return to this point shortly.) However, the ordering given seems to me as natural as any. It is not padded in any way- there is no listing of all CNF's which gives smaller size to the largest CNF.

To learn an unrestricted CNF expression (in the ordering given), Valiant's algorithm can simply be applied starting with $k = 2$ and increasing k until it is successful. For each k, we call $E(k)$ positive examples, for each rejecting as before any k-clauses not containing a literal determined to be true. Here $E(k)$ is the smallest number of examples which give us confidence $1-\epsilon$ that the candidate expression g has in fact learned f, for positive examples. I will quantify $E(k)$ shortly. Having called $E(k)$ positive examples, we call $E'(k)$ negative examples, in order to assure ourselves that g is in fact an adequate representation of f. Here $E'(k)$ is an adequate number of examples so that after $E'(k)$ examples without error, we have confidence $1-\epsilon$ that g will predict future negative examples with accuracy at least $1-\epsilon$. If g mistakes any negative example, we reject g and proceed to the next k. Before k becomes larger than the size l of the largest clause in f, we will find an acceptable expression g.

Naively $E(k)$ would seem to be $L(\epsilon,(2n)^{k+1})$ as in Valiant's algorithm for learning k-CNF's. However we are testing many successive g's, so it might not be so unlikely that one would survive L examples without error even if its probability of error was high. We can correct for this by subjecting each successive g to a more stringent test. If we choose $E(k)$ so that the probability of success when testing k-CNF's is δ^k, then the total probability of accepting an unsatisfactory expression g is less than $\sum_{i=1,\infty}\delta^i$. This is less than ϵ provided we choose $\delta \leq \epsilon-\epsilon^2$. We may achieve this by choosing

$$E(k) = kL(\epsilon-\epsilon^2,(2n)^{k+1}).$$

Since the probability of surviving $L(\delta,(2n)^{k+1})$ examples, each with probability at least δ of removing a clause, is less than δ, the probability of surviving $kL(\delta,(2n)^{k+1})$ is less than δ^k.

Similarly, the naive estimate for E' would be $L(\epsilon,1)$ since $L(\epsilon,1)$ is defined so that in L trials, each with error probability at least ϵ, the probability of no errors is less than ϵ. An adequate choice of E', which corrects for our testing many successive g's, is:

$$E'(k) = kL(\epsilon-\epsilon^2,1).$$

Using this $E(k)$ and $E'(k)$, the total probability that any of our candidate expressions mistakenly be accepted as adequate even though its error probability is greater than δ, is less than $\sum_k \delta^k < \epsilon$. Thus when the algorithm halts, we have confidence $1-\epsilon$ that the CNF g it generates has error probability less than $\delta < \epsilon$.

This algorithm will learn an l-CNF function in time $\sum_{i=1,l} E(i)+E'(i)$ which is $O(l(2n)^{l+1})$. There are, however, $\Omega((2n)^{l+1})$ clauses with fewer than l literals, and thus $\Omega(3^{(2n)^{l+1}})$ l-CNF expressions so T(f) will also be $\Omega((2n)^{l+1})$. I have shown:

Theorem 2: The class of CNF expressions is self-learnable.

Since any Boolean function can be expressed in conjunctive normal form, one might incorrectly assume that Theorem 2 implies that all classes of functions and indeed the class of all functions are learnable. The problem here is that the proof of Theorem 2 only guarantees learning in time polynomial in $T_{CNF}(f)$, where $T_{CNF}(f)$ is the size of f expressed as a CNF (in the ordering given). Some other class of functions F might admit some more compact description, and thus a different T(f). Almost certainly (e.g. if various plausible complexity theory assumptions are true[8]) learnability is not invariant to nonpolynomial changes in the definition of T.

For example, the dual of the algorithm in Theorem 2 for learning unrestricted CNF's, considered to have size given by the ordering stated, will learn unrestricted DNF's, considered again as k-DNF's of all possible k, and ordered so that any DNF with $k+1$ literals in any term has higher order (is larger) than any DNF with no more than k literals in any term. However, I do not know how to learn DNF's ordered as k-term DNF's. Pitt and Valiant[16] in their Theorem 3.7 have observed that any k-term DNF is logically equivalent to some k-CNF, and thus for fixed k, k-term DNF's can be learned by k-CNF's by using Valiant's algorithm. The size of a k-term DNF is $nklog(3)$, since each of k terms can be chosen in 3^n ways. However the size of a k-CNF is about $(2n)^{k+1}$. If the algorithm of my Theorem 2 is applied to learn an n-term DNF, it will not halt in time polynomial in T(f), the size of the n-term DNF.

The class of CNF formulae, which includes all Boolean functions, contains many functions which are essentially random lookup tables. Clearly no algorithm can learn a random lookup table in time polynomial in n and $1/\epsilon$, but independent of T(f). Indeed there are an exponential number of CNF formulae which agree with any set of $p(n)$ randomly classified examples, but which disagree on the classification of a large fraction of the remaining feature vectors.

More generally there is *no* class containing 2^{c^n} distinct functions, for any $c : 1 < c \leq 2$, which is learnable in time $p(n, 1/\epsilon)$ polynomial in n and $1/\epsilon$, but independent of T(f). This is true because $p(n, 1/\epsilon)$ examples can only possibly distinguish $2^{p(n,1/\epsilon)}$ functions, so there will generally be exponentially many functions in F consistent with all the information available to the learning algorithm. I will first present, as an example, a straightforward proof of a result of this type in a special case, and then present a general theorem.

Theorem 3: Let $f'(\vec{x}) = 1$ if $x_1 = 1$ and $f'(\vec{x}) = 0$ if $x_1 = 0$. With every set y of exactly c^n total vectors, associate a function $f_y(\vec{x}) = f'(\vec{x})$ for $\vec{x} \notin y$ and $f_y(\vec{x}) = \bar{f}'(\vec{x})$ for $\vec{x} \in y$. Then the class of functions F=$\{f_y\}$ is not learnable by any algorithm in time bounded by any polynomial $p(n, 1/\epsilon)$, independent of T.

Comment 1: The class F *can* trivially be learned for the special case of a distribution which is uniform over all vectors $\vec{x}$, since the functions f_y only differ on an exponentially small set of vectors. In particular, the learning algorithm: let $g(\vec{x}) = f'(\vec{x})$ (independent of examples) already works unless $\epsilon < (c/2)^n$. But when ϵ is this small, then $(1/\epsilon)^j = 2^n$ for $j = 1/log(2/c)$ and in time polynomial in $1/\epsilon$, one may sample the entire space. That is, as in Valiant's self-learning algorithm for k-CNF's one knows (with probability $1-\epsilon$) that after $L(\epsilon, 2^n) < (1/\epsilon)^{j+2}$ examples, all but a fraction ϵ of the vectors $\vec{x}$ will have been sampled. Thus one may compile a look-up table g in time polynomial in $1/\epsilon$.

Comment 2: This theorem holds on simple information theoretic grounds and is independent of whether P=NP or not.

Proof: Consider the distribution D_z which is uniform over the set z consisting of the first c^n vectors. For any binary function f_b there exists a function f_y such that $f_y|_z = f_b|_z$.

Now any learning algorithm A which sees up to $p(n, 1/\epsilon)$ examples can be regarded as a map from the set of possible classified p-strings to the set of possible functions. In particular A must map each string it sees into one and only one function. (This map may be probabilistic, which will not affect the counting arguments used below.) It is easy to show that there are many more functions then there are strings (even when two functions are regarded as equivalent when they differ on only a fraction ϵ of examples). Thus there must exist many functions which A will get wrong a large fraction of the time.

In fact the number of strings A can see is less than

$$(c^n)^p 2^p = 2^{p(1+nlog(c))}$$

since each of p vectors can be chosen in c^n ways and at most each of p vectors is assigned a 1 or a 0. The number of functions in an ϵ ball around any function on c^n vectors is

$$\binom{c^n}{\epsilon c^n} \le c^{n\epsilon c^n} = 2^{n\epsilon c^n log(c)}.$$

Thus the number of functions, that A can predict within ϵ of, is less than

$$2^{n\epsilon c^n log(c) + p(1+nlog(c))}.$$

However the number of distinct functions of c^n vectors is 2^{c^n}. If I choose n large and $\epsilon < 1/(nlog(c))$, A can predict an exponentially small subset of the possible functions. Thus for any A, there are many functions in the class F which it will get wrong more than a fraction ϵ of the time. **Q.E.D.**

The fact that no class F containing 2^{c^n} distinct functions is learnable, is a simple consequence of some beautiful results of Blumer et. al.[4] which I very briefly review. They say a set of points S is *shattered* by a concept class F if $F|_S$ contains all possible functions on S. The *Vapnik- Chernovenkis (V-C) dimension* of F is then defined as the smallest integer d such that no S of cardinality $d+1$ is shattered by F.

Proposition 3 of [4] states that, if the V-C dimension of F is d, and S contains m points, then the cardinality of $\{F|_S\}$ is no greater than $\Phi_d(m)$ where $\Phi_d(m) = \sum_{i=0}^{d} \binom{m}{i}$ if $m \geq d$, and $\Phi_d(m) = 2^m$ otherwise.

If F contains 2^{c^n} distinct functions on $m = 2^n$ vectors, we thus see that the V-C dimension d of F must be at least $\Omega(c^n/n)$.

Now Theorem 2 of [4] states that no concept class F can be learned without calling at least $\Omega(d)$ examples. This implies:

Theorem 4: No concept class containing 2^{c^n} distinct functions, can be learned in time polynomial in ϵ and n but independent of $T(f)$.

Theorems 3 and 4 followed because polynomially many examples did not contain enough information to solve the desired problem. Similarly Kearns et al.[12] showed their Theorem 15 (that monotone monomials are not learnable using negative examples only by any representation) by a similar technique. When available, such information theoretic arguments are desirable since they do not depend on complexity assumptions such as P≠NP. I now give an algorithm which requires exponential time but is able to learn *any* given class F of functions from polynomially many (in $n, 1/\epsilon$, and $T(f)$) examples. This shows that such information theoretic arguments alone will not be strong enough to establish that any class of functions is not learnable, and suggests that it will be difficult to show that any class of functions is not learnable without assuming (or proving) P≠NP, or perhaps stronger assumptions such as the existence of one way functions[8,13]. This result is akin to showing that the learning problem is in NP in the sense that the problems in NP all admit solution in finite time.

Theorem 2.3 in [3] constitutes a different proof of a similar result.

Theorem 5: *Any* class F of functions may be learned using polynomially many (in $n, 1/\epsilon$, and $T(f)$) examples but large time.

Proof: For any set S of functions, define the function: V_S by $V_S(\vec{x}) = 1$ if and only if for the majority of the functions $s \in S$, $s(\vec{x}) = 1$.

For pedantic clarity consider first the case where F contains fewer than $2^{p(n)}$ distinct functions of n variables, for some polynomial p. The algorithm starts with S=F and candidate $g = V_S$. We call alternately positive and negative examples. If g is wrong on any example, then at least half the functions in S are inconsistent with the example. We then redefine S by removing these functions and proceed with the new $g = V_S$. Let E' be the number of examples so that, if g is successful on E' examples, then we have confidence $1 - \epsilon$ that we have in fact learned f. Before we have seen $E'(p(n) + 1)$ examples, we must have constructed a valid g or we have removed every function in F. As in the proof of Theorem 2 above, we may take

$$E'(i) = iL(\epsilon - \epsilon^2, 1).$$

Here i is the number of candidate functions we have already tested plus one.

Now consider the general case where F may contain up to 2^{2^n} distinct functions. Assume there is some effective ordering of the functions in F and take $T(f)$ equal to the logarithm of the order of f in this list. I will revise the algorithm to learn f using $O(T(f)^2)$ examples. Initialize S as the first 2^n functions, and g as V_S. With no more than nE' examples, we either succeed in learning f, or we eliminate all 2^n functions. In the latter case, let S be the set of the next 2^{n^2} functions and proceed. We then either learn f or eliminate this set of functions after n^2E' examples. If we still have not learned f, proceed by taking S to be the next 2^{n^3} examples, and so on. We must learn f after no more than

$$2nT(f)E'(nT(f)) \leq 4n^2T(f)^2 log(1/(\epsilon - \epsilon^2))/(\epsilon - \epsilon^2)$$

examples, because by then we have reached f and considered the function $g = f$ itself. **Q.E.D.**

§5: Hidden Units and the Transitive Closure of Representations

In the late 1950's and the 1960's a number of workers studied the learning abilities of a simple neural net known as the Perceptron[17,14]. The perceptron, in its simplest form, is a simple linear threshold unit. Defined by a set of weights $\{w_i\}$ and a threshold t, it outputs $v = sgn(\sum_i w_i x_i - t)$, where $sgn(x) = 1$ for $x \geq 0$, and $sgn(x) = 0$ for $x < 0$. The perceptron was devised as a simple abstract model of a biological neuron. Given any set of linearly separable examples, i.e. any set of positive examples $\{\vec{x}_+^\mu\}$ and any set of negative examples $\{\vec{x}_-^\mu\}$ for which there exists a weight vector and a threshold $\vec{w}, t : \forall\mu,\ \vec{x}_+^\mu \cdot \vec{w} \geq t, \vec{x}_-^\mu \cdot \vec{w} < t$, it was proved that the Perceptron learning algorithm converges to correctly classify the examples, i.e. to find such a $\vec{w}$ and t[17,14]. (It is not true, however, that the Perceptron learning algorithm must converge in polynomial time.) Unfortunately, as was stressed by Minsky and Papert[14], many problems of interest are not linearly separable.

A solution to this problem is to extend the class of representations to more sophisticated feedforward neural nets incorporating *hidden units*[19]. Hidden units are threshold units whose inputs may include the features x_i as well as the outputs of other hidden units, but whose outputs are connected either to the inputs of other hidden units or to the input of a distinguished output unit. The classification of a given feature vector is given by the value (1 or 0) of the output unit. A feedforward net of threshold units is defined by a directed acyclic graph G, with n source nodes called inputs and a distinguished sink node corresponding to the output. A weight is associated with every edge and a threshold with every node except the input nodes. When a feature vector $\vec{x}$ is presented at the input nodes (component i to the i-th input) each node computes its value as soon as all the neighbors pointing to it have computed theirs, by simple linear threshold function, i.e. $v_i = sgn(\sum_j w_{ij}v_j - t_i)$, where w_{ij} is the weight of edge (i,j), t_i is the threshold at node i, and the sum is over all edges pointing towards i. The net is thus defined by the graph G, the weights $\{w_{ij}\}$ and the thresholds t_i.

The class of feedforward neural nets may thus be defined as the *transitive closure* of linear threshold functions. The variable set is extended beyond the n feature variables by defining new, dependent variables to be computed as linear threshold functions of the feature variables and previously defined linear threshold functions. If the time required to compute the value of a linear threshold function with i inputs is i, then the time required to compute the output of a feedforward net with no more than M units is less than M^2.

Similarly the transitive closure of any class of concepts C is defined by allowing new dependent variables to be computed as the value of any concept c in the original class C whose arguments are either the feature variables or previously defined dependent variables. If the time required to compute a given primitive concept is τ, the time required to compute a net involving M concepts is no more than $M\tau$.

Back propagation[20,23] is a heuristic algorithm for adjusting the weights and thresholds on a neural net of fixed graph G to learn from examples. Back propagation has had considerable success in practical applications. However Judd[10,11] has recently shown that even the question of whether a fixed net can implement or load a given set of examples is NP-complete. In my opinion, this indicates that to address large problems, we will need to be more flexible. In this paper I consider learning algorithms which have the freedom to vary weights, thresholds, and graph. Using such representations, which will be seen to be complete, the loading or memorization of previously seen examples becomes trivial, and interest is focussed on generalization and prediction. Valiant's formalism is well suited to studying these questions.

§6: Complete Representations

In this section I prove my main result:

Theorem 6: For any class of concepts F and any p-time representation G, if F is learnable by G, then F is learnable by the transitive closure of 3-DNF.

This result is a corollary of Lemma 7, which I will state and prove shortly, and the postulate that any program or algorithm can be represented as a 1 tape Turing machine. I will further appeal to the notion of a *reasonable encoding*, as intuitively defined in Garey and Johnson[7]. The hypothesis is that any *algorithm* can be expressed using any *reasonable encoding* as a deterministic Turing machine, and that the size of the DTM so generated under two reasonable encodings will be polynomially related. Thus any recognition algorithm for classifying positive and negative examples, whatever representation class is used to specify the recognition algorithm, may be concisely written as a DTM.

Lemma 7 implies that any DTM of size N which converges to yield an answer of yes or no in time τ, is equivalent to a formula in the transitive closure of 3-DNF of size polynomial in τ and N. The p-time representation g produced by any learning algorithm A, can be regarded as such a DTM. Thus, given a learning algorithm A which produces in time p(T) a program g in the class G, we may use the explicit mapping given by Lemma 7 to modify A so that A instead produces in time $p'(T)$ a formula in the transitive closure of 3-DNF. This result, and its proof, were suggested to me by Cook's Theorem[5,7] that any instance of any problem in NP can be explicitly mapped to an instance of Satisfiability of polynomially related size. My language and conventions follow those of Garey and Johnson[7].

Although I state Lemma 7 in terms of the transitive closure of 3-DNF's and give an elementary proof, more general results on circuit complexity are of course known, see e.g. [2] and [15]. Thus for any class of functions F, such that the transitive closure of F contains *and, or* and *not* gates, the transitive closure of F will be a complete representation for learning. I will give an explicit example of the argument for the class of feedforward neural nets.

Lemma 7: Let D be a deterministic Turing machine with $|Q|$ states, including a distinguished start state q_0 and two distinguished halt-states q_Y and q_N, with a set $\Gamma = \{0,1,b\}$ of tape symbols, where b is a blank symbol, and a transition function $\delta : (Q - \{q_Y, q_N\}) \times \Gamma \to Q \times \Gamma \times \{-1,+1\}$ (where $\{-1,+1\}$ correspond respectively to translating the read-write head left or right 1 square); which has the property that, initiallized with a tape in which the first n squares contain tape symbol either 0 or 1, and the rest of the tape is blank, it halts in either state q_Y or q_N in time less than or equal τ. Then an equivalent formula in the transitive closure of 3-DNF with no more than $3|Q|(2\tau+1)$ terms may be readily constructed.

Comment: I have encoded the learned recognition algorithm g as a Turing machine which reads the n-dimensional feature vector $\vec{x}$ on the first n places of the tape, and halts in either the yes state or the no state respectively as the instance is a positive or a negative example.

Proof: I must construct a formula in the transitive closure of 3-DNF which has value 1 for vectors $\vec{x}$ which cause D to halt in q_Y and has value 0 otherwise. To do this I will need to construct dependent variables. I will use 3 types of dependent variables. Variable $Q(i,k) : 0 \leq i \leq \tau, 1 \leq k \leq |Q|$ will be intended to be 1 if at time i, D is in state k and zero otherwise. Variable $H(i,j) : 0 \leq i \leq \tau, -\tau \leq j \leq \tau + 1$ will be intended to be 1 if at time i, the read-write head is scanning square j, and 0 otherwise. The read-write head can not scan any squares further than τ squares from its start at square 1 as it only moves one square at a step. Variable $S(i,j,l) : 0 \leq i \leq \tau, -\tau \leq j \leq \tau + 1, 0 \leq l \leq 2$ is intended to be 1 if at time i the contents of square j is symbol l (where symbol $l = 0$ is 0, symbol $l = 1$ is 1, and symbol $l = 2$ is b), and $S(i,j,l)$ is 0 otherwise.

These variables must be expressed as functions of the feature variables, i.e. the variables $x(l) : 1 \leq l \leq n$ which give the values of the symbols in tape squares 1 through l and encode the feature vector, and of course they must be expressed so as to take their intended values. Observe that all the dependent variables at time 0 may readily be so defined. Note that $S(0,j,0) = \bar{x}(j), S(0,j,1) = x(j), S(0,j,2) = 0$ for $1 \leq j \leq n$, and $S(0,j,0) = S(0,j,1) = 0$ and $S(0,j,2) = 1$ for other j. Similarly $Q(0,1) = 1$ and $Q(0,k) = 0$ for other k, i.e. D starts in its start state, and $H(0,1) = 1$, $H(0,j) = 0$ for other j, i.e. D starts with its head at tape square 1.

The values of each variable for other values of i will be defined in terms of the values for $i-1$. Indeed at any given time i, one and only one each of the $Q(i,k)$ and $H(i,j)$ will be 1 and the rest will be 0, and only two $S(i,j,l)$ may be different from $S(i-1,j,l)$. Which are 1 at time i will depend on which are 1 at time $i-1$. The precise specification will be determined by δ, the transition function of D. $\delta : (Q - \{q_Y, q_N\}) \times \Gamma \rightarrow Q \times \Gamma \times \{-1,+1\}$ can be written as a table with $3|Q|$ entries, each of which specifies the next state, head position, and new symbol to be written on the tape. This table is explicitly of form: $Q(i,k) = 1$ if, for any j, for some particular k' and l (where k' and l are given by δ) $Q(i-1,k') = 1, H(i-1,j) = 1$, and $S(i-1,j,l) = 1$; and similar expressions apply for $H(i)$ and $S(i)$. That is to say, each dependent variable at time i can be explicitly written as a 3-DNF in terms of the variables at time $i-1$, where the number of terms in each such 3-DNF is less than $3|Q|(2\tau + 1)$. Thus each dependent variable can in fact be explicitly defined in terms of the feature vector and previously defined dependent variables.

The output of the formula is to be 1 if D halts in q_Y and 0 otherwise. In short *Output* $= Q(\tau, |Q|)$ where state $|Q|$ is q_Y. **Q.E.D.**

To compute the value will require evaluating no more than $\tau(|Q|+4(2\tau+1))$ 3-DNF's, each with fewer than $3|Q|(2\tau+1)$ terms. If $g \in G$ is constructed by A in polynomial time $p(n,T)$, then $|Q| \leq p(n,T)$, and if g halts on any input feature vector in polynomial time $r(n,T)$, then $\tau \leq r(n,T)$. The formula constructed by Lemma 7 will have fewer than $3p(n,T)(2r(n,T)+1)$ terms, and may be computed by evaluating fewer than $r(n,T)(p(n,T)+4(2r(n,T)+1))$ 3-DNF's, each with fewer than $3(p(n,T)(2r(n,T)+1)$ terms.

As stated above, a version of Lemma 7 will hold for the transitive closure of any class of functions which contains *and*, *or*, and *not* gates. For example, we may easily write the 3-DNF formula, given in Lemma 7 as a feedforward neural net. A 3-DNF may be trivially written in terms of linear threshold functions by first implementing each term as a linear threshold function and then implementing the 3-DNF by *or*-ing together the separate terms. A linear threshold function with k weights, each of them 1 and threshold $k-1/2$ gives the *and* of its inputs. A threshold function with k weights, each 1, and threshold 1/2, gives the *or* of its inputs. To compute the Turing machine description given above, for each i, we need to compute the *and* of each possible triple $Q(i,k)H(i,j)S(i,j,l)$, of which there are $3(2\tau+1)|Q|$ possibilities, and also we must compute the $\bar{H}(i,j)S(i,j,l)$ of which there are $3(2\tau+1)$ possibilities. Thus we may construct a feedforward net like that of figure 1 with alternating *and* layers containing $3(2\tau+1)(|Q|+1)$ nodes, and *or* layers, containing $|Q|+4(2\tau+1)$ nodes (one node for each Q, H, and S). There are $2\tau+2$ layers in all and no more than $(2\tau+1)(\tau+1)(4|Q|+7)$ nodes and $(2\tau+1)(\tau+1)(9+21|Q|+3|Q|^2)$ edges in all. Thus we establish:

Corollary 8: For any class of concepts F and any p-time representation G, if F is learnable by G then F is learnable by feedforward neural nets.

Of course there may be a learning algorithm which can learn F on a smaller net than the one generated by A and the embedding I have discussed.

When non-total vectors are included learning becomes a more difficult process as, for many classes of representations, it is NP-hard to determine whether a given non-total vector corresponds to a positive or a negative example[22]. However, the embedding that I have given into the transitive closure of 3-DNF is transparent to this problem- i.e. for any given case it neither improves nor exacerbates this problem. If A generates a representation g which, in polynomial time classifies non-total vectors, the simplest way to encode this as a Turing machine is to allow 4 tape symbols: $\{0,1,*,b\}$. Lemma 7 now goes through smoothly. On the other hand, if we settle for a representation g which converges in p-time on total vectors, and exhaustively check non-total vectors, the same exhaustive check can be performed on the 3-DNF or feedforward neural net.

It is also worth remarking that my methods extend to distribution specific learning results.

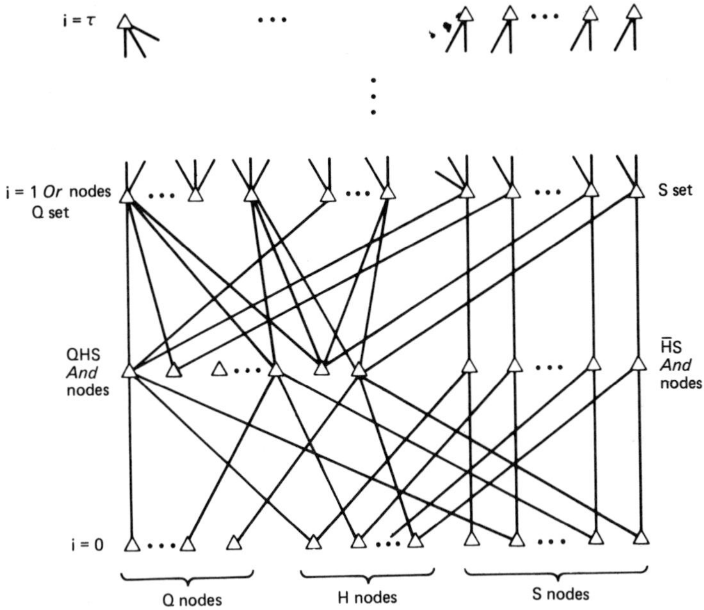

Figure 1: A feedforward net equivalent to the 3-DNF formula of Lemma 7. The net has $2\tau + 1$ layers, alternating between odd layers consisting of a set of Q nodes, a set of H nodes, and a set of S nodes, each a linear threshold function equivalent to the *Or* of several of the node values at the previous layer; and even layers consisting of a set of QHS *And* nodes and a set of $\bar{H}S$ *And* nodes, each a linear threshold function equivalent to the *And* of appropriate nodes at the previous layer. The output of the net is the value of the Q node corresponding to state q_Y at level $i = \tau$.

§7: Discussion

I have shown that feedforward neural nets form a complete representation for learning in the sense that, for any class of functions F and any p-time representation G, if F is learnable by G, then F is learnable by feedforward neural nets. Natural questions which arise now are: (1) to find algorithms which learn interesting classes F on neural nets; and (2) to find classes of functions that are not learnable by neural nets and hence simply can not be learned from examples (in polynomial time). Perhaps the most natural conjecture is that feedforward neural nets themselves form such a class. An answer to this second question would seem to be considerably more interesting than previous results that certain classes of functions are not learnable by particular classes of representations.

Under the complexity theory hypothesis that one-to-one, one way functions exist, Goldreich, Goldwasser, and Micali[8] have constructed classes of functions which are poly-random, that is they can not be distinguished from random functions by any polynomial time test, and are hence unlearnable. Levin[13] has given necessary and sufficient conditions for such poly-random functions to exist. While Theorem 5 indicates it will be difficult to prove any class of functions is not learnable without some complexity theory assumption, it would be interesting to find necessary and sufficient hypothesis for a class of functions to be unlearnable- which would seemingly be a weaker condition than that it be poly-random.

More generally, given the observation that humans somehow grow up to understand the world, it is of interest to characterize broad learning and teaching strategies utilizing complete representations and to characterize learnable and unlearnable classes. By contrast much of the current literature utilizes incomplete representations. Even popular neural network algorithms such as back propagation fix a graph G and only vary weights and thresholds, thus utilizing an incomplete representation. One should perhaps be encouraged by the observation that most negative results on polynomial-time learnability in the literature either depend on a particular incomplete choice of representation[12,16,10,11], or concern function classes explicitly constructed to be unlearnable[8] and thus seemingly unnatural and irrelevant for practical machine learning. On the other hand, no class F containing 2^{c^n} distinct functions, for $1 < c \leq 2$, is learnable in time polynomial in n but independent of T. It is probably not useful in practice to know that any function is learnable in its expression as a CNF, as for most functions this learning will require time exponential in n (although polynomial in T). Thus insight is necessary into what classes are learnable in time polynomial in n and $1/\epsilon$ only and are also useful in describing real world phenomena.

Acknowledgement: I would like to thank Y. Abu-Mostafa for his efforts as editor of this volume, the referee for helpful remarks and particularly for calling my attention to reference [3], and especially L. E. Baum for helpful conversations and critical readings of the manuscript.

References

[1] ANDERSON, D.Z., ed.,*Neural Information Processing Systems- Natural and Synthetic*,

(1988) AIP Press.

[2] BORODIN, A., On relating time and space to size and depth, SIAM J. Comput. V6, N4, (1977) pp733-744.

[3] BLUMER A., EHRENFEUCHT A., HAUSSLER, D., WARMUTH, M. K., Occam's razor, Inf. Proc. Lett. 24 (1987) pp377-380.

[4] BLUMER A., EHRENFEUCHT A., HAUSSLER, D., WARMUTH, M. K., Classifying learnable geometric concepts with the Vapnik-Chervonenkis dimension, Proc. 18th ACM Symp on Theory of Computing, Association for Computing Machinery, New York (1986) pp 273-282.

[5] COOK, S. A., The complexity of theorem-proving procedures, Proc. 3d Ann. ACM Symp. on Theory of Computing, Association for Computing Machinery, New York, (1971), pp151-158.

[6] FORSYTH, R., RADA, R., *Machine Learning Applications in Expert Systems and Information Retrieval*, (1986)Ellis Harwood Limited, Chichester.

[7] GAREY, M. R., JOHNSON, D. S.,*Computers and Intractability: A Guide to the Theory of NP-Completeness*, (1979) W.H. Freeman, New York, pp17-44.

[8] GOLDREICH, O.,GOLDWASSER, S., MICALI, S., How to construct random functions, Proc. 25th IEEE Symp. on Foundations of Computer Science (1984) pp464-479, J. ACM, 33/4 (1986) pp792-807.

[9] GREFENSTETTE, J. J., ed., *Genetic Algorithms and their Applications: Proceedings of the Second International Conference on Genetic Algorithms*, (1987), Lawrence Erlbaum Ass., Hillsdale, N. J.

[10] JUDD, S.,(to appear, Sept 1988) On the complexity of loading shallow neural networks, J. of Complexity.

[11] JUDD, S., (1987), Complexity of connectionist learning with various node functions, COINS tech. report 87-60, U. of Mass, Amherst MA.

[12] KEARNS, M., LI, M., PITT, L., VALIANT, L. G., On the learnability of Boolean formulae, Proc. 19th Ann. ACM Symp. on Theory of Computing, Association for Computing Machinery, New York, (1987), pp285-295.

[13] LEVIN, L. A., One-way functions and Pseudorandom generators, Proc. 17th Symp. on Theory of Computing (1985); Combinatorica 7(4), 1987.

[14] MINSKY, M., PAPERT, S., *Perceptrons, an Introduction to Computational Geometry*, (1969) MIT Press, Cambridge MA.

[15] PIPPENGER, N., FISCHER M. J., Relations among complexity measures, J. Assoc. Comp. Mach. V26, N2 (1979) pp 361-381.

[16] PITT, L., VALIANT, L.G., Computational limitations on learning from examples, Harvard University Tech. Report TR-05-86, (1986).

[17] ROSENBLATT, F., *Principles of Neurodynamics*, Spartan Books, N.Y., (1962).

[18] RUMELHART, D.E., McCLELLAND, J. L.,eds., *Parallel Distributed Processing*, V 1, MIT Press, Cambridge MA, (1986).

[19] RUMELHART, D. E., HINTON, G. E., McCLELLAND, J. L., A general framework for parallel distributed processing, in [18], p48.

[20] RUMELHART, D. E., HINTON, G. E., WILLIAMS, R. J., Learning internal representations by error propagation, in [18], pp 318-362.

[21] TURING, A.M., On computable numbers, with an application to the entscheidungsproblem, Proc. London Math. Soc. Ser 2, v 42, (1936),pp230-265, and v 43, pp544-546.

[22] VALIANT, L. G., A theory of the learnable, Comm. of ACM V27, N11, (1984), pp1134-1142.

[23] WERBOS P., *Beyond Regression: New Tools for Prediction and Analysis in the Behavioral Sciences*, Harvard University Dissertation (1974).

CHAPTER V

Collective Computation, Content-Addressable Memory, And Optimization Problems

J. J. Hopfield

Divisions of Chemistry and Biology
California Institute of Technology
Pasadena, CA 91125

and

AT&T Bell Laboratories
Murray Hill, NJ 07974

A collective decision network is described which can function as a computational element in a digital computer or signal processor. It differs from conventional digital circuit designs in emphasizing the large connectivity and analog response that biological "computational" systems employ. When such circuits have symmetric connections, they display a dynamics which is complex enough to use for computation, but simple enough to yield some general theorems about the dynamics of the circuits. Associative memory and error correcting codes are both computations which fit naturally onto the network dynamics. A large class of computational problems describable as optimizations can be mapped onto such networks. While the networks can be used to make binary decisions, during the decision process they transit the interior of a logical space of which only the boundaries have defined logical meaning.

I Preamble

It is not necessary to know or understand anything about neurobiology in order to pursue the line of analysis to be given. Yet, the "existence theorem" aspects of neural computation are worth keeping in mind as motivation for hardware of an uncoventional type which can do information processing and computation in a less digital fashion. As a signal processor, pattern recognizer, or associative memory, the biological nervous sytems are extraordinarily effective compared to conventional computers. Tasks such as picking out an individual in a crowd are essentially impossible on the largest digital machines, in spite of the fact that the effective "clock rate" of a digital computer is 10^5 to 10^6 times faster than that of a human brain. Part of the success of the biological system is due to the immense parallelism which it uses in tasks such as image processing. The biological system works without a determined detailed circuit diagram in higher animals, for there is no establishable connection between particular brain cells in two different people. The system degrades gracefully, in a fail-soft fashion, when damaged by disease, age, or injury. These are all features which could have interesting uses in man-made computers and algorithms.

Typical mammalian neurons communicate with each other by means of pulses of electrical activity called action potentials. The rate at which such pulses are generated is in general a smooth, sigmoid function of the input to the particular cell. While the accurate timing of individual action potentials is known to be used in hearing and in some other aspects of signal processing, it is believed that information is generally encoded in the rate at which action potentials are generated. This rate ranges between a saturation value of about 500 pulses/second for a cell which is being maximally excited to 0 for a cell which is being inhibited. For many modeling purposes, a nerve cell might be viewed as having a smooth continuous input-output relation as sketched in Fig. 1. It will somtimes be helpful to consider the particular limiting case of this smooth curve as a step-function, though biological neurons or the electrical circuits which might mimic them have continuous responses.

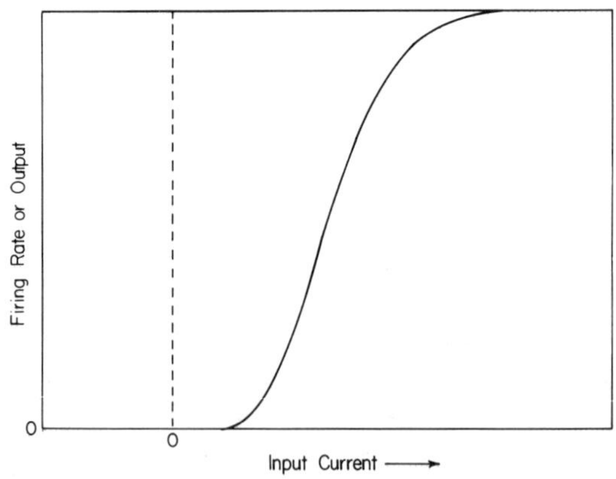

Fig. 1. The effective input-output relation of a typical neuron.

The *numbers* involved in brains are large. The human brain has about 10^{11} nerve cells, each making a complex pattern of connections, with each connected to more than 1000 others. Physical systems having large numbers of components generally exhibit collective phenomena. They display properties which are a consequence of the underlying components and their interactions, but which are very unlike the properties of the components. The interactions between a pair of molecules confined in a box cause simple collisions. When large numbers of molecules are present, these same interactions result in temperature, entropy, phases, turbulence, and the emergent laws of thermodynamics and hydrodynamics. This line of reasoning leads to the idea that there should be collective effects present in brains, that these might be useful in computing, and that some of the unique abilities of a brain might be present also in an electronic technology designed with a view toward collective aspects of neural "hardware".

II Associative Memories and Error Correcting Codes

Associative memory - the kind of memory we have in which one thing reminds us of another - is generally thought of as a very special property of biology [1]. Computer *hardware*, in contrast, contains memory by address. Partial content of a memory is then of no help in finding a particular memory except by exhaustive search, or massive and pre-constructed indexing. We will show how error correcting codes and associative memory are both related to a simple and natural property of a dynamical system.

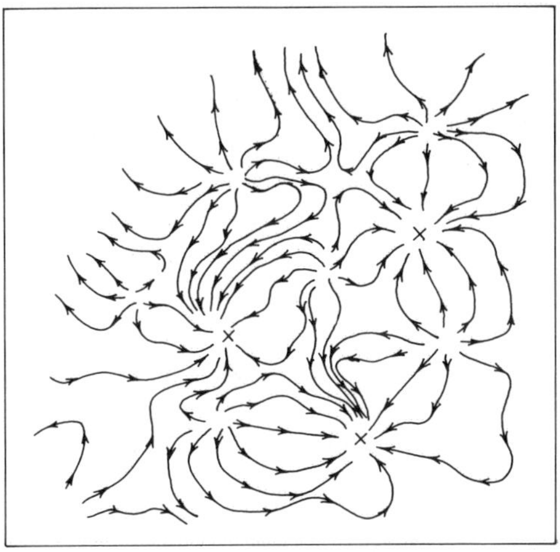

Fig.2. State-space trajectories of associative memory or error correction codes.

Consider an associative memory containing the set of all people you know. In a multi-dimensional information space describing people, each Cartesian coordinate axis could be labeled with a characteristic such as height, weight, name of spouse, or age. In such a space, each person you know is represented by a particular point, whose location describes all characteristics of the individual in question. You know very few of all possible people, so the set of points which represents your friends is very sparse in this space. Suppose now that you are given partial information about some particular friend. This information represents an approximate location in the information space. Presumably this approximate location is closer to the position of this friend in the information space than to the positions of other friends. An associative memory functions by using this closeness to locate or move to the complete information. If a dynamical system is described by a flow in its state space characterized by stable points, as in Fig. 2, the initial information could start the dynamical system at a location close to one of the stored memories. If the state space motion is such that the motion goes to the most nearby stable point, the dynamical system would be functioning as an associative memory. Many dynamical systems display state space motions with multiple stable states. Any such system is a candidate for an associative memory if, in addition, the stable states can be placed wherever they are desired. Thus associative memory is not an esoteric property of computational systems. It is a natural behavior of many simple dynamical systems.

A completely different view can be taken, in which Fig. 2 represents an error-correction code. The final stable states are to be regarded as the code words allowed in some particular error-correction scheme. Other points in this information space represent code words corrupted by noise. The dynamical system then performs the computation of error-correction decoding. Given an initial state corresponding to received information, the dynamical system recovers the code word which is closest to the initial noisy data received. Associative memory is thus not merely memory - it is at the same time a computation or decoding.

The contrast between conventional memory and associative memory as electronic hardware can be seen in the nature of the input and output leads to a memory chip. In an ordinary read-only memory by address (ROM) there are two sets of pins on a chip. Address pins have information sent to them from the computer which specifies the location from which stored information is to be read. Data pins return information to the computer from the addressed location. In an associative memory, however, there is only a single set of pins. *Any* substantial subset of the pins can be chosen for the input information query, and information is then subsequentially read out from all the pins.

Hash-coding and other software constructions have been used to try to make ordinary computer memories somewhat addressable by content. This works acceptably if and only if a particular defined subset of the information will be used as the key, and is quite unlike the democratic access of a truly associative memory. If the hash code is to retrieve the proper memory from information which may contain some errors, difficulties are encountered which can only be solved only by error-correction before retrieving the information. Since this is the same thing as using an associative memory, it does not solve the problem, but rather regenerates the same problem by recursion.

III Collective Associative Memory

Two kinds of collective dynamic systems have been studied for their properties as associative memories and computational devices. The discrete systems use logical output variables, with values 0 and 1 (or -1 and 1). The continuous systems use continuous output variables which have defined upper and lower bounds. The discrete systems are easiest to understand, and illustrate how useful collective properties emerge from large networks.

The continuous systems more nearly represent how a real neurobiological system or electronic circuit having collective properties could function. Under certain limiting circumstances, the stable points of the continuous system are closely related to those of the discrete dynamic systems. We therefore begin with an analysis of the dynamics of the simpler discrete case.

The Discrete Dynamic Algorithm [2]

Consider a set of N "devices", each of which has two states. For simplicity, the two states are described as +1 and -1. The state of the N-device system is described by an N-component vector μ, with components μ_i (for i = 1 to N) or as an N-bit binary word. (Each device is suggestive of a neuron described in the initial section; an output of +1 refers to a maximally activated neuron and -1 to a maximally inhibited neuron.) There is a set of "connections" T_{ij}, from neuron j to neuron i, so that the input to neuron i is given by

$$H_i = input\ to\ i = \sum_{j \neq i} T_{ij} \mu_j + I_i$$

I_i is the specific input to device i from external sources, while the sum is the input to i caused by the connections to the other devices. Self-connections T_{ii} are not allowed.

A dynamic system requires a time evolution algorithm. The time dependence of the state is defined as follows:

a) *pick a device k*

b) *if* $H_k \geq 0$, *set* $\mu_k = +1$

$< 0, \qquad -1$

c) *repeat*

If μ_k already has the correct value, then no change is involved in the setting step. Time is not truly present as it would be in most dynamic systems, but the sequences of states defined by the algorithm are equivalent to a dynamics with one unit of time elapsing between successive applications of the updating algorithm. In step a), the devices k will be picked at random without memory of past choices. (An alternative of picking the devices in a cyclic sequence shares most of the important properties.)

This general system has a dynamics which is too rich in possible behaviors. Such a dynamics could even be used to simulate any digital computer and program. There is, however, one special circumstance which results both in simple and interesting behaviors. When the T matrix is symmetric, and has zeroes on the diagonal then

$$E = -\tfrac{1}{2} \sum_{i,} \sum_{j} T_{ij} \mu_i \mu_j - I_i \mu_i$$

can never increase under the algorithm. The proof is trivial when $T_{ij} = T_{ji}$ and $T_{ii} = 0$. Further, if all the devices are interrogated once and none changes, the state is stable and a local minimum of E. The allowed states of the network are the corners of an N-dimensional hypercube, and the local minimum refers to the fact that all the corners adjacent to the present one have larger values of the "computational energy" E. The existence of this discrete Lyapunov function is a sufficient condition to prove that the only limiting behaviors of the system are stable points, and from any starting point, the system will follow its dynamics to a stable point and cease to change. The dynamics is characterized by a flow like that of Fig. 2.

The stable points of the system are independent of the order in which the devices are sampled in the updating algorithm. If the updating choice is cyclic, the state space flow will be uniquely defined. If the updating is by random choice of the device to update next, then the flow will have a stochastic element, but will always lead downhill and stop. Simultaneous update of many or all devices can have a different character, seen easily by the example on two devices with $T_{12} = T_{21} = +J$

(1)	*(2)*	
I	*-I*	state
-J	*J*	H_i
-I	*I*	next state with simultaneous update
J	*-J*	H_i
I	*-I*	next state with simultaneous update

There has also been much analysis of this system with the deterministic update rule (b) replaced by a non-deterministic one characterized by a temperature, but this generally lies beyond the scope of present discussions [3,4].

In order for this system to function as an associative memory or CAM, it must be possible to place the memories or stable points where they are desired. Suppose several memories μ^s, ($s = 1$ *to* n) are desired. An adequate (but not optimal) T can be chosen as a sum of outer products if the memory vectors are not correlated.

$$T = \sum_s (\bar{\mu}^s \mu^s - I)$$

or

$$T_{ij} = \sum_s \mu_i^s \mu_j^s \qquad i \neq j$$

$$T_{ii} = 0$$

As long as n is not too large, *each* state μ^s will either itself be stable, or there will be a stable state very near (in Hamming distance) to μ^s.

The general thrust of why this should be true can be seen by placing the system in one of the nominal memories μ^t and calculating H_i.

$$H_i = \sum_{j \neq i}^{N} \mu_i^t \mu_j^t \mu_j^t + \sum_{s \neq t}^{n} \sum_{j \neq i}^{N} \mu_i^s \mu_j^s \mu_j^t$$

$$= \mu_i^t (N-1) + (N-1)(n-1) \textit{ terms of magnitude } 1 \textit{ and random sign}$$

The first term is a large "signal" for which the sign of H_i agrees with the sign of μ_i^t, producing stability. The second term is "noise", of mean zero and mean square $(N-1)(n-1)$. Thus, as long as $(n-1)$ is appreciably smaller than $(N-1)$, it is unlikely for the second term to be larger than the first and the state μ^t should be stable.

For $N = 100$, if $n \leq 25$, then the probability of an error in any particular position is about 0.024. The expected number of errors in a memory is then 2.4 bits out of 100. The actual stable points or memories will be very near their nominal locations in Hamming space. For smaller n, the probability of memories being exactly correct grows rapidly. Other more complex algorithms for constructing the T matrix allow the recovery of the precise memories desired without error. The mathematical structure of the stable points in the limit $N \to \infty$, $n \to \infty$, $n/N =$ constant has recently been analyzed [4]. It has some interesting phase transition aspects, and qualitatively corresponds to our elementary analysis until n gets large. The number n of random memories which can be put into memory without error has also been studied [5]. There is a general upper bound on the amount of information which can be reconstructed from using a threshold function (b) and *any* algorithm on such a matrix [6]. From a practical viewpoint, a T matrix has N^2 intersections and can store on the order of $0.1\text{-}0.2N^2$ bits with the elementary algorithm before getting into trouble. Much of the information is in the *sign* of the T_{ij} elements, and T_{ij} can be clipped and still function reasonably, at least if N is not too large.

A less symmetric form of construction is often useful [7-9]. In a representation in which each $\mu_i = 0$ *or* 1, choose

$$T_{ij} = 1 \textit{ if } \sum_{s=1}^{n} \mu_i^s \mu_j^s > 0$$

$$= 0 \textit{ otherwise}$$

Add to the system some global and unmodifiable inhibition, so that the input to i is

$$H_i = \sum T_{ij} \mu_j - \alpha - \beta \sum_{j \neq i} \mu_j$$

Use memories which are dilute in the number of 1's in a memory, with a mean fraction of 1's given by a constant f. If $N = 200$, $f = 0.1$, $\beta = 0.95$ *and* $\alpha = 0$, about 70 memories can be stored at reasonable error levels. Each memory contains

$$200\,(f\ln_2 f + (1-f)\ln_2(1-f)) = 64$$

bits of information, so the total information stored is 4,480 bits. Since the matrix entries are 0 and 1, the matrix contains 40,000 possible bits, so the information storage effectiveness is again about 10%.

The use of dilute memories has several technical advantages. In the symmetrically coded memories, if state $\underline{\mu}^s$ is a stable state, state $-\underline{\mu}^s$, with each entry multiplied by -1, is also stable. In the dilute system the 1's complement of a stable state is *not* a stable state. From the viewpoint of hardware, it is generally easiest to build a system with only one sign of T_{ij} writable, and having only one non-zero value of T_{ij}. The dilute system runs at a level of activity (number of devices "on") which is more like the way biology operates. And finally, it allows the state *00000*..... to be used in a natural way as an indication of no information.

When $\alpha = -0.01$ and $\beta = 0.95$, the memory performs as a forced categorizer. Every initial state, whether it is close to a memory or not, will result in a final memory. An initial state of *100000000*........ is almost equidistant from all memories in Hamming space, and little resembles any. About seven memories will have a 1 in the first position, and one of the seven will be chosen as the stable state. With these values of α and β, the system operates on a scanty similarity.

For $\alpha = 0.3$ and $\beta = 0.8$, a different behavior will occur. Now *10000* ...will lead to the final state *000000...*, which is also stable. The flow map and attractor picture of Fig. 2 is replaced by that of Fig. 3, in which the stored memories are identical to those in Fig. 2, but the flow pattern is different. Now it is necessary to have an appreciable closeness to a particular memory in order to reach that memory, and an initial state which does not much resemble any memory will result in the *000000* state. The behavior is now to recognize and retrieve *only if the initial clue is adequately like a stored pattern.*

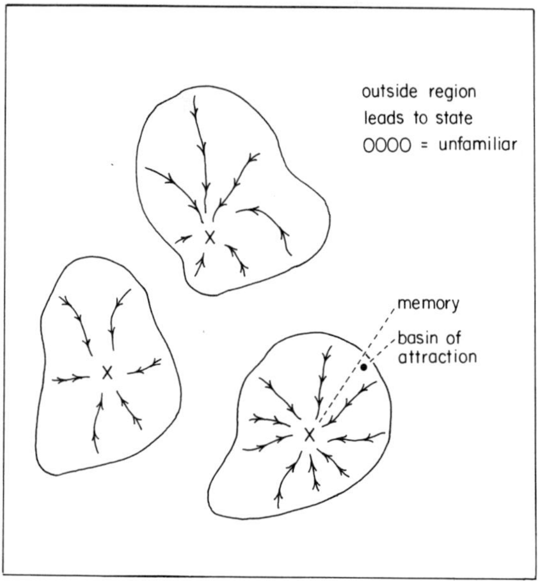

Fig.3. Memory dynamics when adequate similarity is required.

At a higher cognitive level, we all make such adjustments when listening to a speaker. If he speaks our native language, we are precise in our word recognition. A badly accented word spoken in a sentence will fail to be classified, and will not be recognized as a familiar word. When we listen to our language spoken by a speaker with a heavy foreign accent, our behavior is quite different. We will attempt to make any sound into a word, and operate in the forced categorization mode. The same sound rejected earlier will easily be interpreted as the correct word. The difference between these two behaviors is the difference between Fig. 3 and Fig. 2, and can be controlled by appropriate choice of inhibition and thresholds.

In the memory structure of devices, (and perhaps of the brain) a use of such a thresholding would permit an information query to go out in parallel to many physically different parts of a memory. Only the part which found a significance to the information query would turn on and report back information. The rest would simply switch off.

The Continuous Dymanical Algorithm [10]

Consider a set of N saturable amplifiers having input-output relations like that of Fig.- 1. Let each amplifier have an output available of either sign. Connect the output of amplifier j to the input of amplifier i by a resistor R_{ij}. Define T_{ij} by

$$T_{ij} = (\textit{sign of output used at amplifier } j \textit{ by this connection-}) / R_{ij}$$

Each amplifier has an input capacitance C_i and an input resistance ρ_i. The time rate of change of the charge on C_i is given by

$$C_i \frac{du_i}{dt} = -u_i/R_i + \sum_j T_{ij} V_j + I_i \qquad V_i = g_i(u_i)$$

$$R_i^{-1} = \rho_i^{-1} + \sum_j |T_{ij}|$$

R_i is the net resistance to ground seen from the input of amplifier i. g_i is the input-output relation of amplifier i. The circuit diagram envisioned is sketched in Fig. 4. This set of equations describes the time dynamics of the state of the circuit. The instantaneous state of the circuit is defined by the instantaneous values of the voltages V_i. While the description is in terms of amplifiers, a set of biological neurons also plausibly generates this same set of dynamic equations.

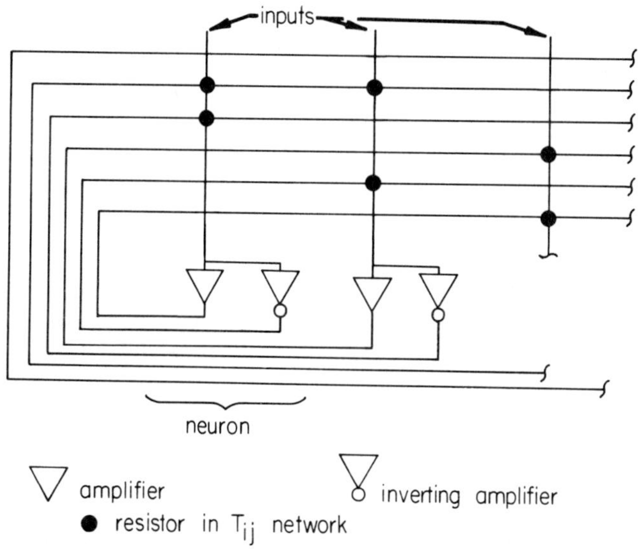

Fig. 4. A set of amplifiers connected as a collective symmetric circuit.

As in the discrete case, there is no general simple description of the behaviors of such a circuit. Again, the special set of cases $T_{ii} = 0$, $T_{ij} = T_{ji}$ leads to simple behavior. Under these circumstances, if g_i is monotonic and bounded, then

$$E = -1/2 \sum_i \sum_j T_{ij} V_i V_j - \sum_i I_i V_i + \sum_i \frac{1}{R_i} \int^{V_i} g_i^{-1}(V) dV$$

then

$$\frac{dE}{dt} \leq 0 \qquad \frac{dE}{dt} = 0 \rightarrow \frac{dVi}{dt} = 0$$

The dynamic system has the computational energy E as a Lyapunov function and a flow of exactly the character of Fig. 2, a flow to stable points or memories. Associative memory and error correction are thus a natural feature of this class of analog circuits.

For simplicity, let the maximum and minimum values of V_i be -1 and 1. The domain of operation of the circuit of N devices is the interior of an N-dimensional hypercube, while the domain of operation of the discrete algorithm was the corners of this same cube.

There is a simple correspondence of memories in the high gain limit. If the g_i are scaled by λ

$$V_i = g_i(u_i) \rightarrow V_i = g_i(\lambda u_i)$$

the integral term in E becomes

$$1/\lambda \sum_i \frac{1}{R_i} \int^{V_i} g_i^{-1}(V) dV$$

This integral term goes to infinity on the boundaries of the hypercube, and restrains the possible values of V_i to the interior of the cube. At the same time, as λ becomes large, the integral term will vanish on the interior of the hybercube. Thus in the high gain limit,

$$E = -1/2 \sum_{ij} \sum T_{ij} V_i V_j - \sum_i I_i V_i \quad \text{on the domain} \quad -1 < V_i < 1$$

Unless T is negative definite, the minima of E will all be on the surface of the hypercube. If $T_{ii} = 0$, the limit points "will be" at the corners of the hypercube which were stable points under the discrete algorithm, neglecting irrelevant special cases. The stable points or memories of a real circuit will be esentially digital in character in the high gain limit, and will be located (and positionable) by the viewpoints developed for the discrete case.

Simulations of the analog equations on a digital computer has shown all the expected results. The system does function appropriately and well as an associative memory in the high gain limit. The stable points are fail soft, generally staying near to the appropriate hypercube corner when some T_{ij} are eliminated. The stable character is not changed even if symmetry is violated to a small degree (for example, by removing some T_{ij} while leaving the corresponding T_{ji} present). The memory is *distributed* in that a given T_{ij} can have information about many memories, and many T_{ij} are involved in the reconstruction of each data bit. The decision as to which memory to retrieve is collective in that the dymanics of retrieval is a parallel and general decision, not a sequence of decisions as would be done in something like a tree search.

IV COMPLEX COMPUTATIONS AND OPTIMIZATIONS

The circuit of Fig. 4 minimizes the computational energy E. Many interesting and difficult computations can be described as optimization problems. Associative memory is itself an optimization problem, involving searching for the memory or code-word which best fits the data. Problems in robotics (what is the optimal set of movements to do a task), routing, and perception can often be stated as optimization problems. When the quanitity to be optimized can readily be mapped into an appropriate energy function, a real collective device could rapidly solve the desired problem, and if not able to find the best example, would find at least a very good one.

The Traveling Salesman Problem has been studied as an example of how to put a complicated combinationally difficult problem onto such a "neural" or amplifier network [11]. In this problem, a set of n cities $A, B, C...$ are given, with distance d_{AB} between the cities. The problem is to find the path going to each city once and returning to the starting point such that the total distance D traversed

$$D = \sum_{path} d_{AB}$$

is a minimum.

There are three aspects to this problem.

- We must make a syntactical agreement with the network about how we will interpret its final state. This is a necessary part of using any computer. If a computer prints its output as a sequence of bits on a tape, we must know whether to interpret the output as a long binary number, as a string of ASCII characters, or in some other format. This also defines the internal representation of the path in the computing machine.

- There must be appropriate terms in E which serve to allow only answers (stable states) of the agreed-on syntax. While for some problems any configuration has an interpretation, in general there will be many states which are not answers. These states are not *incorrect* answers, they are *disallowed* as answers since they are not interpretable. The syntax terms in E must prevent such disallowed answers. Again, this problem arises in ordinary programming. If, for example, the computer is supposed to print out an English sentence as its answer, the program must use English syntax rules to prevent outputs which are strings of words but not sentences. (The trivial solution of explicitly storing all allowed answers in memory is prohibited when the number of possible answers is exponential in the size of the problem.)

- A data term in E must correspond to the information about what is to be minimized in the particular problem. The expectation is that the best (deepest) minimum will have the greatest accessability and be most easily found.

The E-function defines the connections and inputs in the real network. The dynamics of the network then minimizes the data term subject to the syntax restrictions.

In the n-city traveling salesman problem, $N = n^2$ amplifiers are arranged in an $n \times n$ matrix, and the columns labeled by an integer *1,2...* and the rows by the names of the cities *A,B,C....* The amplifiers will have minimum outputs of *0* and maximum outputs of *1*. The syntax will be chosen so that in the final stable state, one amplifier in each row and each column will have an output very near *1*, and the rest will have outputs of *0*. Thus, the acceptable final states will be permutation matrices. We will interpret the answer as a statement of the order in which to visit the cities. In the final state displayed in Fig. 5, the interpretation is "start at ***A*** go to ***E***, then ***G*** .... and return to ***A***.

A computational energy ***E*** which is appropriate to the TSP problem is

$$E = A/2 \sum_X \sum_i \sum_{j \neq i} V_{Xj} V_{Xi} + B/2 \sum_i \sum_X \sum_{Y \neq X} V_{Xi} V_{Yi}$$

$$+ C/2\Big(\sum_X \sum_i V_{Xi} - n\Big)^2 + 1/2 \sum_X \sum_{Y \neq X} \sum_i d_{AB} V_{Xi} (V_{Y,i+1} + V_{Y,i-1})$$

The first three terms generate the correct syntax, and yield zero computational energy if and only if the final state is a permutation matrix. The first two terms penalizes having more than one *"1"* in any row and column, while the third term penalizes any matrix which does not have n entries which are *"1"*. The data term is equally simple, for it has a numerical value of $+D$ for a tour of length D.

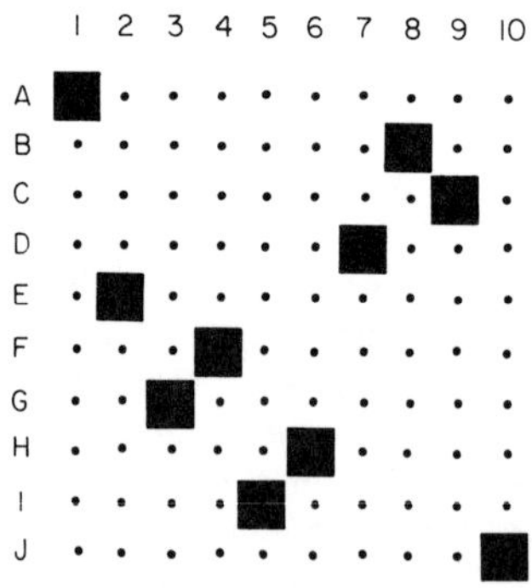

Fig. 5. A typical final state of the Traveling Salesman network. The linear dimension of a square is proportional to the output of an amplifier.

There is $2n$-fold degenerate description of a particular closed state path in this energy function, corresponding to the n different cities which could be chosen for the start and the 2 directions in which a closed path may be taken. Because of this degeneracy, the system is started from a state of $V_i = 0.1 +$ random noise. The random noise is necessary to break the degeneracy. Otherwise, the network would not choose a particular one of these $2n$ representations of a given path.

Of course, the network does not know about E. It follows a dynamics which decreases E, rather than one which find a global minimum. Thus, it is not guaranteed to find the best path. The E function has a large number of local minima. The suggestion that the network might perform well comes from the idea that the good solutions are the lowest E points, and the size of an attractive basin should generally be a very rapid function of the energy of the bottom of the basin. There is no theory of how well the network should do, so we studied the problem empirically.

For $n = 10$, the network is very effective. There are about 2×10^5 different paths, and the network found one of the best two paths about 50% of the time. For $n = 30$, simulations were much slower, and much less data was acquired. There are about 10^{30} paths, and the paths found lay among the best 10^6 - 10^7 of these paths. The bias in favor of good paths is a factor of 10^{23} or so.

The analog nature of the system is essential to obtaining good answers. If the gain is too large, the system finds answers little better than random. Of course, the gain also cannot be chosen too small, or else the 1 and 0 aspect of the final state will not occur. When the gain is correctly chosen, the computation is made by a trajectory on the interior of the state space, and the trajectory reaches a corner of the state space late in the convergence. A single collective, global decision is made. This is illustrated in Fig. 6. In this figure, the output of each neuron is indicated by the linear dimensions of the square in the appropriate location. In this particular convergence, city J was held in position 10 to eliminate most of the degeneracy.

This intermediate state is not a 'logical' position. Each amplifier can be regarded as corresponding to a proposition. For example, if amplifier B3 is a *"1"*, the proposition that city *B* should be visited third is true, while if B3 is a *"0"*, this proposition is false. The intermediate state shown in Fig. 6 has *partial* amplitudes for *contradictory* propositions, such as B8, C8, and C9. This non-logical state is, however, reasonable. It corresponds to the idea that city C is near position 8 in the tour, though whether it is in points 6, 7, or 8 is not yet defined. The right hand half of the figure shows the shortest path for the particular set of locations used. In other words, the positions of the cities on the tour appear to be *roughly* defined early in the calculation, and this rough definition is refined a

Many other combinatorial problems can be put onto such networks by similar techniques. In some cases the data is in connections T, and in other cases the data can be put into inputs I. These problems include graph-in-subgraph, graph coloring, Euclidean match, Hamiltonian circuit, and some kinds of decoding problems. (The *mathematics* of codes based on choosing subsets of bits which are mutually odd (or even) maps onto such networks easily in a mathematical sense, by using higher order energy terms. Thus, a term

$$\mu_1 \mu_2 \mu_3 \mu_4 \mu_5$$

would have energy -*1* if the five bits are mutualy odd, and *+1* if the bits are mutually even. But the physical implementation of this mathematics is difficult compared to the case of quadratic energy functions.

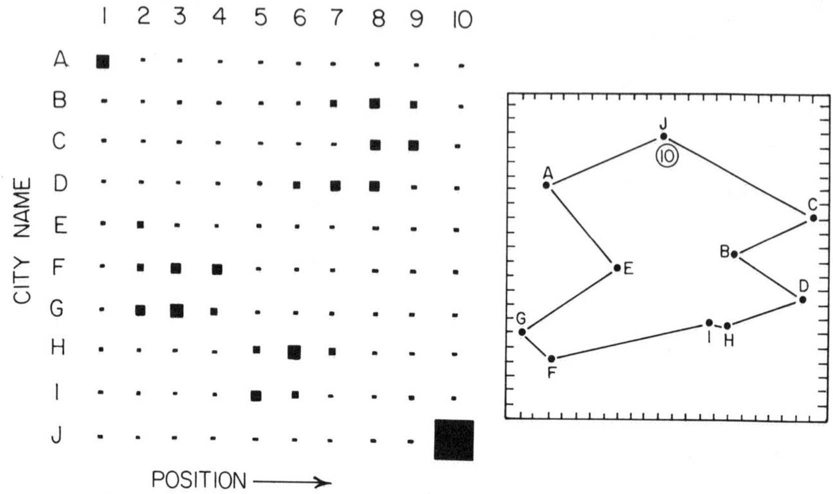

Fig. 6. An intermediate stage during the convergence of the network to the path shown at the right.

Because the circuits potentially perform a rapid analog convergence to the final stable state, and can be used in such a fashion that the output is compatible with digital processing, efforts are underway to fabricate hardware and examine its real computing possibilities. Such efforts have included the building of macroscopic circuits from standard components, [5,12] implementation in optics, [13] fabrication in an analog VLSI mode, [14] and research on making high density matrices of resistive components using techniques outside those used in standard VLSI [12,15]. Potential uses for such hardware range from massive associative memories to front-end signal processing of high-data rate signals such as those present in speech and vision. They are particularly interesting to applications where massive amounts of soft or imprecise data must be evaluated (and thus the analog limits to the precision of the hardware is not an intrinsic disadvantage), and where so much computation needs to be done that the amount of computation per device or transistor becomes important. The circuits would be extraordinarily effective in device terms. While in the TSP problem the best solution will not be found, in VLSI a very good solution could be found using very few gates and within a sub-microsecond time scale.

In the general area of decoding, such circuits offer some interesting possibilities. Using them in an associative memory mode, brute force decoding using stored code words could be feasible on a very large scale. Analog decoding of at least some codes appears possible. And computational decoding of cleverly chosen codes, in which small networks have huge numbers of *orderly* stable states, may also be possible.

The relation of the analog domain in which the networks calculate or converge to the discrete space of combinatorial problems is tantalizing. One is reminded of the use of the interior of a space in linear programming, even though it is known that the actual solutions will be at surface points. The relation of the discrete system to the spin glass [16] provides a connection between the mathematics of statistical mechanics and computational issues.

These networks are too simple and too unstructured, and the "neurons" to unsophisticated to be viewed as a literal interpretation of what biology does. They are at best a crude description of how a very small part of a brain might behave in a short time scale. But they do provide at least a metaphor for how the use of the major features of biological neurons - an analog and sigmoid response, large connectivity, and reciprocal connections - lead naturally to fast, powerful, novel, and fail-soft collective computational abilities.

I thank D. W. Tank for ongoing discussions of these matters.

REFERENCES

1. G.E. Hinton and J.A. Anderson, eds., *Parallel Models of Associative Memory,* Lawrence Erlbaum Assoaciates, Hillside, NJ (1981).

2. J.J. Hopfield, *Proc. Natl. Acad. Sci. USA* **79**, 2554 (1982).

3. G.E. Hinton and T.J. Sejnowski, *Proceedings of the IEEE Conference on Computer Vision and Pattern Recognition*, 448 (1983).

4. D.J. Amit,, H. Gutfreund, and H. Sompolinsky, *Phys. Rev. Lett.* **55,** 1530, (1985).

5. R.J. McEliece and E. C. Posner (preprint).

6. Y. S. Abu-Mostafa and J. St. Jacques, *International Symposium on Information Theory* (Abstracts) IEEE Catalog No. 85 CH 2201-2, 74 (1985).

7. J. Lambe, A. Moopenn, and A. P. Thakoor, *Am. Inst. of Aeronautics and Astronautics,* **85-5007,** 160 (1985).

8. D. J. Willshaw, Models of Distributed Associated Memory (thesis), Univ. of Edinburgh (1971).

9. A. E. Gelperin, J.J. Hopfield, and D.W. Tank, in *Model Neural Networks and Behavior*, A. Selverston, ed., Plenum Press, New York (1985).

10. J. J. Hopfield, *Proc. Natl. Acad. Sci. USA,* **81**, 3088 (1984).

11. J.J. Hopfield and D.W. Tank, *Biol Cybernetics* **52**, 141 (1985).

12. J. Lambe, A. P. Thakor, and A. Moopen, JPL Technical Report D-2825, Jet Propulsion Laboratory, Caltech, Pasadena, CA (1985).

13. D. Psaltis and N. Farhat, *Optics Letters* **10**, 98 (1985).

14. M. Sivilotti, M. Emmerling , and C.A. Mead, 1985 *Conference on Very Large Scale Integration*, H. Fuchs, ed., Computer Science Press, Rockville, MD (1985), p. 329.

15. L.D, Jackel, R.E. Howard, H.P. Graf, B. Straughn, and J.S. Denker, *J. Vac. Sci. Tech.* (in press).

16. F. Tanaka and S.F. Edwards, *J. Physics F. Metals Phys.* **10**, 2769 (1980); G. Toulouse, *Helv. Phys. Acta*, **57**, 459 (1984); P. Peretto, *Biol. Cybernetics* **50**, 51 (1984).

CHAPTER VI

Complexity of Random Problems

Y. S. Abu-Mostafa

Electrical Engineering and Computer Science Departments
California Institute of Technology
Pasadena, CA 91125

The notion of random problems has been introduced to capture the difficulty of modeling natural pattern recognition problems. These problems do not have a concise effective definition. We discuss the heuristic and formal aspects of random problems, and quantify their difficulty through a number of complexity measures. The relations between these measures provide lower and upper bounds on the cost of the system that solves a given random problem. Finally, we address the role of learning by example in solving random problems from a theoretical and practical points of view.

I. Introduction

It is evidently very difficult to model natural environments such as visual scenes. In certain applications, such as commercial data compression, simplified models seem to be adequate for the job. However, in applications such as pattern recognition, simplified models invariably lead to unsatisfactory results when the method based on the model is applied to realistic inputs.

The difficulty in modeling natural environments is reflected in the failure to find satisfactory answers to some very basic questions about these enivronments. One such question is: what makes two pictures 'similar'? Answering this question amounts to defining a distance metric on the class of pictures, such that similar pictures will be a small distance apart. Failure to define such a metric is the reason why a theory as elegant as rate-distortion theory has not had a notable practical impact in speech communication for example.

Is it possible to solve the pattern recognition problem in a natural environment without having a good model for the environment in question? Let us consider for example the problem of recognizing the presence of a tree in a picture (a 512×512 array of bits). It may seem possible to perform this task without knowing what a tree is, simply by writing an algorithm that takes the 512×512-bit input array and performs some preprocessing, edge detection, and parsing to decide whether there is tree or not. However, a closer look reveals that the algorithm must have the definition or the model of a tree embedded in it, for one can in principle use the algorithm to classify every possible 512×512 array into tree/no tree thus exactly defining what it means to be a tree in a picture.

The necessity for having, explicitly or implicitly, a model for the environment is particular to pattern recognition. In a different application such as computer graphics, one may be able to generate what looks like a very natural tree without having, explicitly or implicitly, a model for what a tree is. The reason is that the graphics algorithm will be considered successful if it can produce a number of pictures containing real-looking trees, not *all* pictures containing real-looking trees. One can therefore focus on a very small subset of trees which have an underlying regularity that makes them easier to generate but does not take away from their natural appearance. The problem with pattern recognition is that the algorithm should be ready for an *arbitrary* input.

Is the difficulty in finding good models for natural environments inherent? Unless and until a model is found, this remains a subjective question since we do not know, in formal terms, what these environments are (if we knew, that would be the model!). This difficulty extends to problems of pattern recognition in natural environments. We will study a formal class of problems called random problems, which are inherently difficult to model. The definition of random problems formalizes the plausible properties of natural pattern recognition problems. The relevance of this chapter to natural pattern recognition problems is thus a thesis, not a provable fact. The notion of random problems can also be studied as a mathematical notion, without regard to its possible connections with pattern recognition.

The organization of this chapter is as follows. We motivate and formalize what it means for a problem to be random in section II. This includes a dichotomy between random problems and structured problems. In section III, we introduce different measures that quantify the complexity of a random problem, derive basic relations between these complexity measures, and discuss other measures that can be defined. The real complexity, however, lies in the automated learning of a random problem, which is the subject of section IV. This section discusses the basic issues involved in a learning process.

II. Random Problems

We will consider the decision problem as defined by a function $f:\{0,1\}^N \to \{0,1\}$ where N is arbitrary but fixed. For example, in the above discussion of recognizing the presence of a tree in a picture, the problem would be formalized as a function $f:\{0,1\}^{512\times 512} \to \{0,1\}$, where $f(\mathbf{x})=1$ iff $\mathbf{x}$ (interpreted as a 512×512 visual scene) represents 'a picture that contains a tree'. Whereas natural patterns are usually represented by analog values on a continuous domain, we shall deal with patterns represented by a finite number of bits. This is adequate for conveying all the main ideas, and generalization to analog representations can be made using rate-distortion methods.

As a computational problem, the tree-in-a-picture problem as well as many other natural pattern recognition problems is best posed as a finite problem (fixed N). This contrasts the canonical way of posing computational problems that allows the input to be any one of an infinite number of possible instances (variable N). For example, the Hamilton-circuit problem (does a given graph have a 'path' of edges that contains every vertex once and only once) can be posed for any graph of any size. The canonical way allowing variable N is necessary to invoke the tools of algorithms and time complexity which require uniformity with respect to N. Posing the tree-in-a-picture problem as a uniform problem would require the extension of the finite problem (say with $N=(512)^2$) to arbitrarily large values of N. However, there is no natural extension for the tree-in-a-picture problem that parallels conventional problems such as the Hamilton-circuit problem. For example, having finer sampling of the scene (say $N=(1024)^2$) does not really provide a 'bigger' instance of the problem, just a better (and perhaps more informative) representation of the same instance.

Fortunately, our characterization of problems as random versus structured will not require uniformity with respect to N. Therefore, we will not be forced into artificial extensions of finite problems. Let us develop the notion of random versus structured for problems of the form $f:\{0,1\}^N \to \{0,1\}$. Consider the following two problems where $N=30$ and the input string $\mathbf{x}\in\{0,1\}^{30}$ is interpreted as the binary representation of a number in the range from 0 to $2^{30}-1 \approx 10^9$. The first problem addresses the question 'is this number a prime?', i.e., $f(\mathbf{x})=1$ iff $\mathbf{x}$ represents a prime number. To solve this familiar problem, one can apply any of the known tests for primality to the input $\mathbf{x}$. Alternatively, one can take advantage of the finite nature of the problem and consult an exhaustive list of all primes in the range of values $\mathbf{x}$ can assume.

The second problem is less familiar and offers fewer choices for solution. Given $\mathbf{x}$, it addresses the question 'is this the social security number of a convict?', i.e., $f(\mathbf{x})=1$ iff $\mathbf{x}$ represents the social security number of a convict.

There is clearly a basic difference between the 'prime' problem and the 'convict' problem. Let us first rid the 'convict' problem from secondary attributes that may confuse the issue. The notion of a convict may be a bit ambiguous, but one can clearly make it well defined by spelling out some details such as 'convicted by a US court of law'. Even with the notion of a convict unambiguously defined, the basic difference between the two computational problems persists. This difference is what distinguishes structured problems from random problems.

In order to solve the 'convict' problem, we need a list of the social security numbers of all convicts. Although we can also solve the 'prime' problem by consulting the list of primes, we have the option there to use a primality test. The primality test can be viewed as a short effective definition of the notion 'prime'. In the case of the notion 'convict', it is highly unlikely that there is a simple relation between the social security number and the legal status of a person. Therefore, the list of convicts which constitutes an effective definition of the notion 'convict' cannot be compressed into a significantly shorter definition. Problems which do not have a concise effective definition are called *random problems* [1,3]. The characterization 'random' is based on the Kolmogorov complexity approach to randomness [9], which is the lack of a concise effective definition. It is unrelated to the probabilistic connotation of the word 'random'. This will be made precise shortly.

Having no concise effective definition does not mean that the list of social security numbers of convicts is the shortest effective definition of the notion 'convict'. Indeed, there may well be simple relations that allow the list to be somewhat compressed. This property is typical in many natural pattern recognition problems that can be modeled as random problems. Invariably, there will be some regularities that can be taken advantage of. These regularities often obscure the random nature of these problems. An effective definition of a problem has to cover the 'irregular component' as well, and may therefore have to be long even if there are many regularities in the problem.

To formalize the notion of a random problem, we use the Universal Turing Machine [14,17] as our computational model. Let U be a fixed Universal Turing Machine. U can be thought of as a standard general-purpose computer that runs programs on data in the usual way. Let $f:\{0,1\}^N \to \{0,1\}$ be the problem in question. Computing f on the machine U requires a program $\mathbf{p} \in \{0,1\}^*$ (a finite binary string which can be thought of as the compiled program in machine language). The length of the program $\mathbf{p}$, denoted by $|\mathbf{p}|$ in bits, is the main criterion here. If f can be computed with a short program, f will be considered a structured problem. If the shortest program that can compute f is longer than a certain threshold, f will be considered a random problem.

What does it mean for a program $\mathbf{p}$ to compute f on the machine U? There are two plausible definitions which turn out to be essentially equivalent, as far as the length of the program is concerned. We may require that U, starting with $\mathbf{p}$ and any input $\mathbf{x} \in \{0,1\}^N$, terminates leaving the value of $f(\mathbf{x})$ (this operation is denoted by $U(\mathbf{p},\mathbf{x}) = f(\mathbf{x})$). This definition corresponds to the situation where we have successfully written a pattern recognition program that is ready to take any input pattern and classify it correctly. The other definition requires **p** to define f, rather than to compute it for a given input. Let $\tau(f)$ be a listing of the truth table of f, i.e., $\tau(f) = \tau_0\tau_1 \cdots \tau_{2^N-1}$ where τ_k is the value of $f(\mathbf{x})$ when $\mathbf{x}$ is the N-bit binary representation of the number k (thus $\tau(f)$ is the 'last column' in the truth table of f that holds the 2^N binary values $f(\mathbf{x})$ assumes as $\mathbf{x}$ varies over $\{0,1\}^N$). Defining f amounts to specifying $\tau(f)$, and we thus require that U, starting with **p**, terminates leaving $\tau(f)$ as output. In symbols, $U(\mathbf{p}) = \tau(f)$. In this definition, there is no particular input we want to classify. However, both definitions imply that **p** has complete information about f. In the first definition, this is shown by the ability of **p** to compute $f(\mathbf{x})$ for an arbitrary **x**, while in the second definition this is shown by the fact that **p** provides the entire truth table of f.

The reason why these two definitions are essentially equivalent in terms of the length $|\mathbf{p}|$ is that each program can be simply modified into the other without significant increase in length. If we have the first program that takes **x** and computes $f(\mathbf{x})$, we can modify it by adding a routine that generates all $\mathbf{x} \in \{0,1\}^N$, lexicographically, runs the program on each **x**, and saves the computed value of $f(\mathbf{x})$. The result is that the modified program ends up producing the entire truth table $\tau(f)$ which is what the second program is supposed to do. On the other hand, if we have the second program that produces $\tau(f)$ without reference to any particular input **x**, we can modify it by adding a routine that takes a particular **x**, runs **p** to generate $\tau(f)$, and then searches through $\tau(f)$ for τ_k that corresponds to $f(\mathbf{x})$. When it finds this τ_k, it erases everything else and terminates. The result is that the modified program ends up producing the value of $f(\mathbf{x})$ for a particular **x** which is what the first program is supposed to do. It is easy to verify that these modifications lengthen each program by only $o(N)$ bits. We choose the second program **p** that generates $\tau(f)$ and base our formal definitions on it.

There are many different programs **p** that can generate the truth table $\tau(f)$ for a given problem $f:\{0,1\}^N \Rightarrow \{0,1\}$, and some of these programs will be shorter than others. The shortest possible length of a program **p** that generates $\tau(f)$ is called the Kolmogorov complexity [9] of the string $\tau(f)$. In general, the Kolmogorov complexity of a string $\mathbf{s} \in \{0,1\}^*$ is defined by

$$K(\mathbf{s}) = \min \{ \ |\mathbf{p}| \ \ | \ \ U(\mathbf{p}) = \mathbf{s} \}$$

(again, $U(\mathbf{p}) = \mathbf{s}$ denotes that U, starting with **p**, terminates leaving **s**). The Kolmogorov complexity of **s** measures the degree of randomness or lack of structure in **s**. The more random strings **s** will have larger values of $K(\mathbf{s})$. However, $K(\mathbf{s})$ will never have to be larger than $\approx |\mathbf{s}|$, for one can always generate **s** by a program **p** whose function is 'print **s**', and this program is only marginally longer than **s**. In fact, the most random strings **s** do have $K(\mathbf{s}) \approx |\mathbf{s}|$, which can be shown by counting the number of short programs and realizing that there are not enough of them to generate all strings of length $|\mathbf{s}|$. When applied to $\mathbf{s} = \tau(f)$, the Kolmogorov complexity leads to the following definition [2]:

Definition. The *randomness* R of a problem $f:\{0,1\}^N \to \{0,1\}$ is defined by

$$R(f) = \log_2 K(\tau(f)) \quad \text{bits}$$

The logarithm is taken to normalize $R(f)$ to the range from ≈ 0 (where $\mathbf{p}$ is very short, hence $\tau(f)$ is highly structured) to $\approx N$ (where $\mathbf{p}$ is as long as $\tau(f)$ itself, hence $\tau(f)$ has no pattern whatsoever).

The value of $R(f)$ is what determines whether f is a structured problem or a random problem. The idea of the definition of random problems is that they have larger values of $R(f)$ than do structured problems. There is a number of technical variations on how one can approach this definition. We will fix a threshold R_o (the particular choice of which is not crucial to most of the theory) and introduce the following definition [3].

Definition. A problem $f:\{0,1\}^N \to \{0,1\}$ is said to be *random* if $R(f) \geq R_o$.

For example, if our threshold $R_o=20$, we will call a problem random if it has no effective definition (i.e., if there is no program that generates $\tau(f)$) which is shorter than 2^{20} (about a million) bits. One can hardly expect a human programmer to be able to write a program for such a problem. It is not the program length, per se, that makes us expect that human programmers will not be able to write the program. After all, programming teams have written extremely elaborate programs in a systematic way. It is rather the fact that the *shortest* program for the problem is very long that implies that there is no pattern in the problem that can be understood and translated systematically into computer code by a human programmer.

With $R_o=20$, the vast majority of classification problems that can be defined on a 512×512 image are (very) random, but most of them have no practical significance. Indeed, most of the conventional computational problems are structured, not random. One example is the Hamilton-circuit problem. Let us take $N=(512)^2$ and interpret the input $\mathbf{x}$ as a 512×512 adjacency matrix of a graph on 512 vertices, where 1 in the (i,j) position means that there is an edge between vertex i and vertex j, and 0 means that there is no edge. To compute $f:\{0,1\}^{512\times512} \to \{0,1\}$ is to determine whether there is a path of edges in the graph that contains every vertex once and only once. One algorithm to do that is to exhaust all possible (cyclic) permutations of the vertices and check, for each permutation, if there is an edge between every consecutive pair of vertices. Coding this algorithm as a program does not take very many bits (although running the program will take a very long time). Whenever there is a short algorithm for a problem, no matter how inefficient, the problem is structured. In spite of being a structured problem, the Hamilton-circuit problem is believed to be an intractable problem, requiring time exponential in N to be solved. It is the length of the program, not the running time, that makes a problem random or structured.

The use of the Universal Turing Machine U as the computational model on which the definition of random problems is based has certain ramifications. On the one hand, it makes the definition quite general since any other 'reasonable' model of computation can be simulated by U (the Church-Turing thesis) without increasing the length of the program involved by more than a constant. On the other hand, U may be too powerful a model for practical purposes. For instance, U is allowed to run the program **p** to generate $\tau(f)$ for any length of time as long as it eventually terminates. It is a standard result in the theory of computation that some programs will run for an exceedingly long time before they terminate. It seems of no practical use to have a short program that would generate $\tau(f)$ if it were allowed to run for millions of years! One remedy for this problem is to introduce a less general but more practical definition of randomness that includes a restriction on the running time. One such definition requires a program **p** that computes $f(\mathbf{x})$ for any $\mathbf{x} \in \{0,1\}^N$ within T steps, where T is some practical time function of N. Let $U_T(\mathbf{p}) = \mathbf{s}$ denote that U, starting with **p**, terminates within T steps leaving **s** as output. The time-bounded randomness can be defined by

$$R_T(f) = \log_2 \min \{ \ |\mathbf{p}| \ \mid U_T(\mathbf{p},\mathbf{x}) = f(\mathbf{x}) \}$$

and a definition of time-bounded-random problems can be based on that. Another dilemma is resolved when such restricted definitions of randomness are adopted. This logical dilemma arises when we want to *prove* that a given problem is random. Chaitin's version of Gödel's incompletenesss theorem [6] implies that it is impossible to prove that a given problem is random (using the usual axiomatic systems, and assuming R_0 is not very small). However, it becomes possible (though perhaps difficult) to prove that a given problem is time-bounded random when an effective time bound T is used. It is readily possible, however, to prove that *most* of the problems in some simple classes are random problems.

The difficulty in proving randomness for specific problems creates an awkward situation for the application of the theory. When we consider a specific problem such as the tree-in-a-picture problem or the Hamilton-circuit problem, we may be able to tell when it is a structured problem by finding a simple algorithm for it, but we will be unable to tell with certainty when it is a random problem. The uncertainty is apparent when we are dealing with a natural pattern recognition problem for which no simple algorithm has been found. Should we assert that such a problem is a random problem? There are compeling, yet inconclusive, informal arguments for and against the assertion. However, the mere fact that the problem may be structured is of no use if the 'short program' cannot be found. In practice, problems which appear to be random, but may have an unknown wry structure, will be treated as random problems. Fortunately, the approach for solving random problems which is discussed in section IV, will still work if the problem turns out to be structured. In section III, we will prove that certain problems are random, with high probability.

Another practical modification in the definition of randomness aims at tolerating small errors. If it is considered too rigid to require **p** to produce the truth table of f perfectly (or equivalently classify each input **x** correctly), one can relax this requirement. Assumung a probability distribution on the input space $\{0,1\}^N$ (measuring relative frequency or relative importance of different **x**'s), and allowing error δ of the time, the following definition emerges.

$$R_\delta(f) = \min \{ R(g) \mid \Pr(f(\mathbf{x}) \neq g(\mathbf{x})) \leq \delta \}$$

where, instead of insisting on computing $f(\mathbf{x})$ or $\tau(f)$, we settle for a function g which does not differ much from f, but is simpler to describe. If $R_\delta(f)$ is used instead of $R(f)$ to define random problems, some previously random problems will not be considered random any more because they have a structured approximation. Using $R_\delta(f)$ may be more plausible practically, not only because in practice some error can usually be tolerated, but also because natural pattern recognition problems are typically fuzzy and it would be an overspecification to require perfect classification of every input. For instance, some images equally qualify as containing a tree or not containing one, depending on who is looking. More importantly, some 'images' do not look like pictures of anything in the first place and we may not care how our tree-in-a-picture algorithm would classify them. A score of other variations of $R(f)$, and a corresponding number of notions of a random problem, can be produced by merging the error tolerance, the bound on computation time, and other practical considerations.

III. Complexity Measures

In this section, we address the difficulty of random problems through a number of complexity measures. These measures are based on different models of computation, ranging from simple lists to Turing machines. The values of these measures reflect the difficulty of the problem from the point of view of the respective models. Our first complexity measure is the randomness $R(f)$ itself, which is the basis for categorizing a problem $f:\{0,1\}^N \longrightarrow \{0,1\}$ as random $(R(f) \geq R_o)$ or structured. Although this categorization is binary (random versus structured), the actual value of $R(f)$ describes in more detail how random or how structured a problem is, and how big the system that implements f has to be. We start by investigating the distribution of values $R(f)$ assumes.

Let us take N sufficiently larger than R_o, say $N > R_o + 10$. Let f be any of the 2^{2^N} problems that can be defined with N Boolean inputs. The range of values of $R(f)$ is determined by $R(f) = \log_2 K(\tau(f))$. The Kolmogorov complexity $K(\mathbf{s})$ of any string $\mathbf{s}$ ranges from the minimum length of any syntactically correct program, to the length of the program 'print $\mathbf{s}$'. Therefore, depending on the technical details of the Universal Turing Machine model U, there are constants C_1 and C_2 such that

$$C_1 \leq K(\mathbf{s}) \leq |\mathbf{s}| + C_2$$

Since $|\tau(f)| = 2^N$, $R(f)$ ranges from ≈ 0 to $\approx N$ bits. In fact, the vast majority of the 2^{2^N} problems $f:\{0,1\}^N \longrightarrow \{0,1\}$ have $R(f) \approx N$. To see this, we estimate the number of problems having $R(f) < N-1$. Each f in this class must have a distinct program $\mathbf{p}$ of length $|\mathbf{p}| < 2^{N-1}$ that generates $\tau(f)$. However there are at most $2^0+2^1+2^2+ \cdots +2^{2^{N-1}-1} < 2^{2^{N-1}}$ programs in this length range, which is much less than 2^{2^N}. Hence, almost all problems are crowded around $R(f) \approx N$. This also implies that almost all problems are random problems.

We can use this fact to construct a random problem in a probabilistic manner. We construct $\tau(f)=\tau_0\tau_1 \cdots \tau_{2^N-1}$ by flipping a fair coin independently to determine each τ_k. Since this experiment generates any of the 2^{2^N} problems with equal probability, and since almost all of these problems are random problems, the experiment will, with overwhelming probability, generate a random problem. This experiment highlights the relation between the notion of randomness in the Kolmogorov sense as used in defining random problems and the notion of randomness as used in probability theory. If we generate a problem in a random (probability) way, we will probably get a random problem (Kolmogorov)! It is important not to confuse the two notions. However, it is useful to use the fact that probability is involved in constructing a problem to draw conclusions about its likely status as a random problem.

The second complexity measure is based on the observation that, in many natural pattern recognition problems, the truth table $\tau(f)$ has many more 0's than 1's, or vice versa. For example, in the tree-in-a-picture problem, the set of pictures containing trees is a tiny subset of the set of all 512×512 binary arrays. $\tau(f)$ of this problem has far fewer 1's than 0's. The reason for this common situation in natural pattern recognition problems is that the input is usually represented in a highly redundant fashion. This means that many allowable input patterns never occur in practice (correspond to *don't care* [8]). When the representation is redundant, the dimensionality N of the problem is more than the 'essential' dimensionality. Our second complexity measure quantifies this essential dimensionality, or entropy [2].

Definition. Let $h(f) = \min\{\, |f^{-1}(1)|\,,\, |f^{-1}(0)|\,\}$. The (deterministic) *entropy* of f is defined by

$$H(f) = \log_2(1 + h(f)) \quad \text{bits}$$

where $f^{-1}(1) = \{\mathbf{x} \mid f(\mathbf{x})=1\}$ and $f^{-1}(0) = \{\mathbf{x} \mid f(\mathbf{x})=0\}$, which means that $h(f)$ is the number of 1's or the number of 0's in $\tau(f)$, whichever is smaller. Clearly $0 \leq h(f) \leq 2^{N-1}$, hence the entropy $H(f)$ ranges from 0 to $\approx N$ bits, essentially the same range as $R(f)$.

In some sense, $H(f)$ is a weak complexity measure. Knowing $H(f)$ alone does not provide much information about the complexity of f. For instance, a function with few 1's (hence low entropy) may be fairly complex if the 1's are scattered at random in the truth table, whereas a function as simple as $f(\mathbf{x})=x_1$ (the first bit of $\mathbf{x}$) has the maximum possible entropy. However, in the Appendix, we show $R(f)$ can be at most $\approx H(f)$, hence the entropy serves as an upper bound for the randomness.

We have used the symbol '$\approx$' without formal definition. What we mean by $\approx A$ is $A \pm o(N)$, where $o(N)$ is a function of N which is asymptotically negligible with respect to N. The error terms represented by $o(N)$ are a consequence of formalization technicalities and simulation overhead, and they will vary depending on specific details of the models.

The strength or weakness of a complexity measure is a consequence of the strength or weakness of the model on which the measure is based. The criterion in each measure is the size of the specification of $\tau(f)$ using the model. For example, $R(f)$ is based on the Universal

Turing Machine and the criterion is the length of the program $\mathbf{p}$ that runs on U to generate $\tau(f)$. Similarly, we can think of $H(f)$ as being based on a model that genrates $\tau(f)$ when we specify where the 1's are (or the 0's, whichever are fewer). The size of the specification is taken as the number of points we need to specify. Because the model is simplistic, the size of the specification is sometimes more than necessary because it does not take advantage of any relation between where the 1's (or 0's) are in $\tau(f)$. The Universal Turing Machine, on the other hand, is a powerful model that can take advantage of very subtle relations and this allows $\mathbf{p}$ to be shorter and shorter.

What makes a model more powerful than another? If model A can simulate model B, but not vice versa, A is more powerful than B. Simulating B means taking a description of B together with the input to B and simulating what B would do on the input. For example, the Universal Turing Machine can be programmed in a straightforward manner to generate $\tau(f)$ from the specification of where the 1's (or the 0's) are, while the simplistic model that does only that cannot simulate an arbitrary program running on the Universal Turing Machine. This is why the randomness $R(f)$ is less than or equal to $\approx H(f)$. In fact, these two measures are extreme in that their models are very weak or very strong. Other measures based on intermediate models will lie between $R(f)$ and $H(f)$.

The third complexity measure is based on a model which is a tradeoff between model power and practical realization. The model is combinational circuits [14], which are loop-free interconnections of logic gates. Combinational circuits cover a wide spectrum of practical electronic circuits as well as certain classes of neural networks. The measure is called $C(f)$, the *circuit complexity*. It measures, on a logarithmic scale, the minimum size of a circuit implementation of the problem f. The range of $C(f)$ is also $0 \leq C(f) \leq N$ bits. The particular choice of the logic gates in the circuit that defines $C(f)$ affects the value of the measure only marginally. In the Appendix, we give a concrete definition for $C(f)$, and show that $C(f)$ lies between $R(f)$ and $H(f)$, i.e., $R(f)$ is at most $\approx C(f)$ and $C(f)$ is at most $\approx H(f)$. If we neglect terms of the order $o(N)$, we can write

$$R(f) \leq C(f) \leq H(f)$$

This relation will also hold for other variations of $C(f)$ which are not based on a purely combinational model. In practice, when we try to build systems to solve pattern recognition problems, we do not have all the details of the system at the outset. The above relation provides bounds on one important detail, how big the system has to be.

While $R(f)$ is as small as a measure can be that is based on a 'realizable' model, $H(f)$ is as large as a measure can be that is based on a 'useful' model. Measures based on other practical models with varying degree of sophistication will lie somewhere between $R(f)$ and $H(f)$. $C(f)$ is such a measure, and it is practically more relevant to the cost of implementing f than $R(f)$ is, although $R(f)$ is more profound.

The differences between these complexity measures are also meaningful quantities. For example, $H(f)-R(f)$ is called the *false entropy* of f. Since $H(f)$ does not take advantage of any regularity in $\tau(f)$ while $R(f)$ takes advantage of all effective regularities, false entropy quantifies the regular part or the hidden structure of the problem. For example, the tree-in-a-picture problem has a significant false entropy due to such regularities as shift and scale invariance in visual scenes and a number of other, more subtle, properties. The identification of these regularities dramatically simplifies the problem, since in effect it reduces the entropy by the corresponding number of bits. Since the actual size is exponential in the complexity measure, one bit difference cuts the system size by half. We will come back to false entropy when we discuss learning in section IV.

Finally, we define a version of $C(f)$ that accommodates some error tolerance, in the same way we defined $R_\delta(f)$ in section II. We assume a probability distribution on the set of inputs $\{0,1\}^N$ and define

$$C_\delta(f) = \min \{ C(g) \mid \Pr(f(\mathbf{x}) \neq g(\mathbf{x})) \leq \delta \}$$

where δ is the allowed probability of error. Clearly, $C_\delta(f) \leq C(f)$. In words, $C_\delta(f)$ measures the smallest size of a circuit that can approximately solve the problem f.

IV. Learning

When a problem $f:\{0,1\}^N \rightarrow \{0,1\}$ is declared random, the implication is that no small system, be it an algorithm or a circuit, will be able to solve it. This constraint, by itself, does not pose a serious difficulty for solving the problem in practice (for example, if $C(f)=30$, this implies a circuit with about a billion gates). Indeed, recent and projected technological advancements make it possible to build huge and intricate systems in a routine fashion. The difficulty lies in *finding* the system that solves a given a problem. The shortest program to solve a random problem will necessarily have no logical pattern to it; if it did one could take advantage of this pattern and write a shorter program. The practical impossibility of finding a near-optimal system purely by 'thinking about it' is the premise for introducing the notion of a random problem in the first place. If the premise proves false, the distinction between random problems and structured problems will be that of degree, not of nature.

However, if we do not insist on a near-optimal system, there is a canonical solution to any problem, be it random or structured. We store $\tau(f)$ in memeory and when we want to classify an input $\mathbf{x}$, we search through $\tau(f)$ for the corresponding τ_k. This may be a viable solution for the 'convict' problem for example, where $\tau(f)$ is not too long. We may even take advantage of the low-entropy nature of the problem and store only the $\mathbf{x}$'s where $f(\mathbf{x})=1$ (the list of convicts), and search through this list when we want to classify an input. However, if we want to solve the tree-in-a-picture problem, this approach is totally out of the question. Neither N (the relevant parameter for the storage of $\tau(f)$) nor $H(f)$ (the relevant parameter for the storage of the 1's of f) is small enough to make these exhaustive schemes possible.

Example. Consider a simplified tree-in-a-picture problem $f:\{0,1\}^{512\times512} \rightarrow \{0,1\}$ constructed as follows. We take one particular picture containing a tree (say a picture you took in your back yard and digitized into 512×512 binary pixels), call it $\hat{\mathbf{x}}$. We then pick 100 fixed pixels scattered around (uniformly if you will) and designate these as noise pixels that are allowed to be 1 or 0. Having at most 100 scattered 'spots' out of more than 250,000 pixels will not obscure the appearance of the picture. The problem is thus defined by $f(\mathbf{x}) = 1$ if, and only if, $\mathbf{x}$ agrees with $\hat{\mathbf{x}}$ in all pixels except (possibly) the designated noise pixels. Solving this problem by the above exhaustive schemes requires prohibitive storage. Indeed, storing the 1's of this extremely simplified pattern recognition problem requires more than 10^{30} bits, which will remain prohibitive even when this book becomes old!

The difference between an exhaustive solution and an optimal solution is quantified precisely by the false entropy. We pay an exponential price for every bit of false entropy we fail to 'remove'. This reflects the essential difficulty in solving pattern recognition problems. Exhaustive schemes cannot be implemented, and non-exhaustive schemes cannot be found. While the former difficulty is unsurmountable, the latter may be overcome if we can find an *automated* method for constructing the system, rather than try to construct the system ourselves. The method of learning is a good candidate. Learning is the mechanical process that takes as input pieces of information about f and produces as output an implementation of f. For example, in real life, children see and hear about trees, then become able to recognize trees. The following questions address some key issues pertaining to a learning process.

1. Is the input information about f sufficient?
2. How costly, and how good, is the output implementation of f ?
3. Is the learning process feasible?

Example. One of the simplest models of a learning process is an algorithm L whose input is the complete information about f, namely $\tau(f)$, and whose desired output is a program $\mathbf{p}$ of the shortest possible length that can generate $\tau(f)$. However, such an algorithm does not exist. This fact is a consequence of the uncomputability of the Kolmogorov complexity [9].

This example implies that the automatic removal of the entire false entropy is impossible. When the desired output is a combinational circuit rather than a program, the uncomputability question goes away since one can in principle find the smallest circuit by exhaustive search. The feasibility of learning in these cases becomes a question of how much time is needed to produce the output. Valiant [18] and others formalize this question and give partial answers in a number of cases. Unfortunately, most of the results about interesting learning tasks are negative, predicting that the learning problem is intractable. However, this research area is still at its infancy, and a lot of work needs to be done before solid conclusions can be drawn about the feasibility of automated learning for practical problems. In what follows, we will discuss some basic considerations.

The basic learning scheme is *learning by example.* In this scheme, the learning process is provided with a number of examples $\mathbf{x} \in \{0,1\}^N$ together with the corresponding values $f(\mathbf{x})$. The process is supposed to use this information to produce a system that implements f for all inputs. Learning by example in this pure form reflects the ideal situation where we do not have to do any thinking, just pump examples into the process and wait for the system to pop out! As one would expect, experimental results of this scheme are not very promising. A closer look reveals inherent limitations. There must be restrictions on the class of problems from which f is chosen if a (non-exhaustive) set of examples will suffice to define f to the learning process. Moreover, how the process extrapolates the set of examples must match the restrictions of the class of problems in question. Given all that, the question remains about how much time is needed for the process to produce the system.

How the process extrapolates a set of examples to the entire input space is a major consideration in learning which is called *generalization.* Any viable learning process must have some capability of generalization. We cannot hope to provide an exhaustive set of examples in a practical situation, not only because of their astronomical number, but also because of the difficulty of producing them (e.g., the set of all pictures containing trees). Generalization forces the assumption of underlying regularity, and if the correct generalization is made, this amounts to removing false entropy. Any number of questions can be asked about tradeoffs between the parameters of a generalization process.

It is generally a waste of resources (sometimes fatally so) to depend entirely on a set of examples to tell the process about f when we can use our partial understanding of the problem to give 'hints' to the process to point it in the right direction. For example, the learning process of the tree-in-a-picture problem will treat scaled versions of the same picture as unrelated pictures if we do not tell it somehow about the scale invariance of visual scenes. Some hints are given to the process in a subtle way. When we provide a given process with a non-exhaustive set of examples, we implicitly hint (perhaps falsely!) that the way the process naturally generalizes is the right way. Some major hints are included in the choice of representation of the input. A visual scene represented by $(512)^2$ pixels with no adjacency information is a hopeless starting point for learning. Each bit of regularity missed by the process doubles the number of apparently independent cases it has to worry about. It is a challenge for theory and practice alike to combine our partial understanding of the problem with learning by example to produce a successful pattern recognition system.

Appendix

In this appendix, we provide a technical definition for the circuit complexity $C(f)$ and prove the relations between $R(f)$, $H(f)$, and $C(f)$. Our building block of circuits is the universal gate. A universal gate with n inputs and 1 output simulates an arbitrary Boolean function of n variables. A combinational circuit is made up of any number of universal gates (usually with different n's) interconnected in a loop-free manner.

Definition. The (denormalized) *cost* of a universal gate of n inputs ($n \geq 0$) is defined to be 2^n 'cells'. The cost of a collection of gates is the sum of the costs of the gates.

This definition is motivated by the actual number of cells in an electronic memory, and by the fact that implementing an n-input universal gate requires an exponential number of standard gates. A collection of Q universal gates with $n_1, \cdots, n_Q$ inputs costs $\sum_{i=1}^{Q} 2^{n_i}$ cells. We use these gates together with the input lines $x_1, \cdots, x_N$ to build a combinational circuit Γ to simulate a given function. Γ simulates f if f is one of the gate outputs $y_1, \cdots, y_Q$.

Definition. The (normalized) *circuit complexity* of a Boolean function f, denoted by $C(f)$, is defined by:

$$C(f) = \log_2 \min \left\{ \text{cost of } \Gamma : \Gamma \text{ simulates } f \right\} \quad \text{bits}$$

A constant function f is the output of a universal gate with zero inputs. Such a gate costs $2^0 = 1$ cell. Hence, $C(f) = 0$ bits for the two constant functions, and $C(f) > 0$ for all other functions. Also, $C(f) \leq N$ for any function f which depends on N variables, since a universal gate with N inputs (2^N cells) can simulate any such function. Notice that $C(f)$ differs by at most a constant from the normalized circuit complexity based on any other complete basis of switching devices such as 2-input NAND gates. To see this, one can simulate universal gates using the complete basis and vice versa. For example, a 2-input NAND gate can be simulated by a 2-input universal gate which costs 4 cells. Hence, the simulation costs 4 times the number of NAND gates, in cells. When the logarithm is taken, the two measures differ by at most a constant. Similarly, an n-input universal gate that costs 2^n cells can be simulated by $A \times 2^n / n \leq A \times 2^n$ NAND gates [16].

We now prove [2] that $C(f) \leq H(f) + o(N)$. Functions of entropy H have $h = 2^H - 1$ 1's or 0's in their truth tables. We are interested in estimating the circuit complexity of the functions of entropy H. Without loss of generality, we shall consider only the functions with h 1's.

Definition. Given a function f of entropy H, the state of the variables in a subset S of $\{x_1, \cdots, x_N\}$ is said to be *positive* if there is an assignment of 0's and 1's to the *rest* of the variables that makes $f(\mathbf{x}) = 1$.

f can have at most $h = 2^H - 1$ positive states for any subset S, since there are only h 1's in the truth table of the function. Therefore, as far as f is concerned, the state of the variables in S can be encoded using $\lceil \log_2 (1 + h) \rceil = \lceil H \rceil$ binary variables (the extra 1 represents 'the state is not positive'). Taking $|S| > \lceil H \rceil$, this encoding constitutes information compression, since we represent a number of variables by a smaller number of variables. Furthermore, in terms of the new variables (the compressed variables from S together with the rest of the variables outside S), the entropy of the function remains the same, and hence we can repeat this compression.

Consider an arbitrary function f of N variables whose entropy is H bits. Since we can compress any number of variables into $\lceil H \rceil$ variables, we repeatedly compress $\lceil H \rceil+1$ variables into $\lceil H \rceil$ variables, each time using $\lceil H \rceil$ universal gates of $\lceil H \rceil+1$ inputs. We thus reduce the N variables to $N-1$, $N-2$, ... , down to any number of variables, say $\lceil H \rceil+1$ variables. We can then implement the function f in terms of these $\lceil H \rceil+1$ variables using one universal gate of $\lceil H \rceil+1$ inputs. The compression from N to $\lceil H \rceil+1$ variables takes $(N-\lceil H \rceil-1)\times \lceil H \rceil$ universal gates of $\lceil H \rceil+1$ inputs each, in addition to the final gate with $\lceil H \rceil+1$ inputs. Therefore, the cost of this circuit is $(1+(N-\lceil H \rceil-1)\times \lceil H \rceil)\, 2^{\lceil H \rceil+1}$ cells. Since $0 \le \lceil H \rceil \le N$, this cost is $\le 2^{H+o(N)}$ cells. This proves that $C(f) \le H(f)+o(N)$.

In order to prove that $R(f) \le C(f)+o(N)$ (hence $R(f) \le H(f)+o(N)$), we need to estimate the distribution of $C(f)$, which is done in the following proposition.

Proposition. Let $\hat{N}_K = |\ \{ f : \{0,1\}^N \rightarrow \{0,1\} \mid C(f) \le K \}\ |$. For $0 \le K \le N$, the following holds:

$$\log_2 \log_2 \hat{N}_K \ \le\ K + \log_2 (6+N)$$

Proof. We can take $K \ge 1$ since the statement is clear for $K < 1$ (constant functions only). We will use $\exp(x)$ to denote 2^x when convenient. Let $M = \lceil K \rceil$. We overestimate $\hat{N}_K$ by $\hat{N}_M$. To do this, we shall estimate the number of different ways we can choose a collection of gates given the total cost of 2^M cells, the number of circuits that can be formed using a given collection of gates, and the number of functions that can be simulated on a given circuit. We restrict the gates to have a positive number of inputs, since zero-input gates contribute only the constant functions which can be simulated otherwise. Restricting the cost to be *exactly* 2^M cells is justified by adding 1-input gates (all costs involved are even) without using their outputs.

1. A collection of gates is isomorphic to a multiset of numbers (a multiset is a set where repetition of elements is allowed), where each number corresponds to the number of inputs in a gate. The number of multisets $<n_1, \cdots, n_Q>$ with positive n_i's satisfying $\sum_{i=1}^{Q} 2^{n_i} = 2^M$ (i.e., whose cost is 2^M cells) is at most 2^{2^M}. To see this, let ρ_M be the number of different multisets $X_M = <n_1, \cdots, n_Q>$ where the n_i's are positive integers satisfying $\sum_{i=1}^{Q} 2^{n_i} = 2^M$ for the positive integer M. We claim that any X_{M+1} can be written as the union of two X_M's, except $X_{M+1} = <M+1>$. This can be proved by taking the X_{M+1} and replacing each two 1's in it by a single 2 (conceivably leaving a single 1 at the end). We next replace each two 2's by a single 3 and continue in this fashion until we replace the $M-1$'s by M's. This procedure must yield exactly $<M, M>$ because there is at most a single 1, a single 2 , ... , a single $M-1$ left and these cannot contribute to $\sum_{i=1}^{Q} 2^{n_i}$ more than $2^M - 2$. Therefore, we go back and decompose the two M's getting two X_M's which proves the claim. Hence ρ_{M+1} is at most $1 + \dfrac{\rho_M(\rho_M+1)}{2}$. For

$\rho_M \geq 2$, which holds for $M \geq 2$, this is at most ρ_M^2. The proof now follows by induction after overestimating ρ_1 and ρ_2 by 2^{2^1} and 2^{2^2} respectively.

2. Given N variables $x_1, \cdots, x_N$ and calling the outputs of the Q gates $y_1, \cdots, y_Q$, we have at most $N+Q$ different variables that can be input to each gate. Therefore, for each collection of gates with $n_1, \cdots, n_Q$ inputs, we have at most $\prod_{i=1}^{Q} (N+Q)^{n_i}$ possible interconnection schemes or circuits. This can be rewritten as $\exp\left(\sum_{i=1}^{Q} n_i \log_2 (N+Q) \right)$. Subject to $\sum_{i=1}^{Q} 2^{n_i} = 2^M$, the maximum value of $\sum_{i=1}^{Q} n_i$ occurs when all the n_i's are 1's and is 2^{M-1}. Since Q is also at most 2^{M-1}, we get the overestimate $\exp\left(2^{M-1} \log_2 (N + 2^{M-1}) \right)$.

3. For each of the above circuits, there are $\prod_{i=1}^{Q} 2^{2^{n_i}}$ ways to program the universal gates and, for each of these, we have at most Q implemented functions out of the Q gates. Hence, the number of functions that can be implemented on the circuit is at most $Q \prod_{i=1}^{Q} \exp 2^{n_i}$. This is at most $\exp\left((M-1) + 2^M \right) \leq \exp\left(2\times 2^M \right)$.

Therefore, from 1,2, and 3, the number of functions f that can be implemented within cost 2^M cells is at most the product of the three estimates, namely $\exp((3 + \frac{1}{2} \log_2 (N + 2^{M-1}))2^M)$. Since $M < K+1$ and also $M \leq N$, this is at most $\exp((6 + \log_2 (N + 2^{N-1}))2^K)$. Since $N \leq 2^{N-1}$ (N is an integer), this is at most $\exp((6 + \log_2 (2^{N-1} + 2^{N-1}))2^K)$, which reduces to $\exp \exp (K + \log_2 (6+N))$. |

Thus, the number of functions with $C(f) \leq K$ is at most $2^{2^{K+o(N)}}$. A corollary of the proposition is that almost all functions of N variables have $C(f) \geq N - o(N)$, which is a known result [16].

We can now prove that $R(f) \leq C(f) + o(N)$ by constructing a program to generate $\tau(f)$ based on the smallest circuit that simulates f. We fix a lexicographic ordering of *all* circuits, with the less costly circuits coming first. The program $|\mathbf{p}|$ includes the smallest index of a circuit that simulates f (by the proposition, this index will be at most $\log_2 2^{2^{C(f)+o(N)}} = 2^{C(f)+o(N)}$ bits long), a constant-length routine to 'decode' the circuit from its index, and a constant-length routine to generate $\tau(f)$ from the circuit. The details are straightforward.

References

[1] Y. Abu-Mostafa, "Complexity of random problems," *Abstracts of Papers, IEEE International Symposium on Information Theory*, Brighton, England, IEEE Catalog # 85 CH 2201-2, p. 84, June 1985.

[2] Y. Abu-Mostafa, "The complexity of information extraction," *IEEE Trans. on Information Theory*, vol. IT-32, pp. 513-525, 1986.

[3] Y. Abu-Mostafa, "Random Problems," *Journal of Complexity*, Academic Press, vol. 4, 1988.

[4] A. Aho et al., *The Design and Analysis of Computer Algorithms,* Addison-Wesley, 1974.

[5] A. Borodin, "On relating time and space to size and depth," *SIAM J. on Computing*, vol. 6, pp. 733-744, 1977.

[6] G. Chaitin, "Gödel's theorem and information," *International Journal of Theoretical Physics,* vol. 22, pp. 941-954, 1982.

[7] R. Duda and P. Hart, *Pattern Classification and Scene Analysis,* Wiley-Interscience, 1973.

[8] Z. Kohavi, *Switching and Finite Automata Theory,* McGraw-Hill, 1978.

[9] A. Kolmogorov, "Three approaches for defining the concept of information quantity," *Information Transmission*, vol. 1, pp. 3-11, 1965.

[10] O. Lupanov, "Complexity of formula realization of logical algebra," *Problems of Cybernetics*, vol. 3, pp. 782-811, 1960.

[11] P. Martin-Löf, "The definition of random sequences," *Information and Control*, vol. 9, pp. 602-619, 1966.

[12] C. Mead and L. Conway, *Introduction to VLSI Systems,* Addison-Wesley, 1980.

[13] N. Pippenger, "Information theory and the complexity of Boolean functions," *Mathematical Systems Theory*, vol. 10, pp. 129-167, 1977.

[14] J. Savage, *The Complexity of Computing*, Wiley-Interscience, 1976.

[15] C. Shannon, "A mathematical theory of communication," *Bell Sys. Tech. J.,* vol. 27, pp. 379-423, 1948.

[16] C. Shannon, "The synthesis of two-terminal switching circuits," *Bell Sys. Tech. J.*, vol. 28, pp. 59-98, 1949.

[17] A. Turing, "On computable numbers with an application to the Entscheidungsproblem," *Proc. London Math. Society*, vol. 42, pp. 230-265, 1936.

[18] L. Valiant, "A theory of the learnable," *Communications of the ACM*, vol. 27, pp. 1134-1142, 1984.